動物繁殖

張響英，楊曉志 主編

ANIMAL REPRODUCTION

目 錄

專案一　採精 ··· 1
Project Ⅰ　Semen Collection ·· 1

　任務　採精 ··· 9
　Task　Semen Collection ··· 9

專案二　精液的處理 ·· 19
Project Ⅱ　Semen Treatment ··· 19

　任務 1　精液的品質評定 ··· 35
　Task 1　Evaluation of Semen ··· 35

　任務 2　精液的稀釋 ·· 44
　Task 2　Dilution of Semen ·· 44

　任務 3　冷凍精液的製作 ··· 46
　Task 3　Production of Frozen Semen ··· 46

　任務 4　精液的保存與運輸 ··· 49
　Task 4　Preservation and Transportation of Semen ························ 49

專案三　發情鑑定與輸精 ·· 54
Project Ⅲ　Estrus Identification and Insemination ························ 54

　任務 1　母牛的發情鑑定與輸精 ·· 73
　Task 1　Estrus Identification and Insemination of Cows ·················· 73

　任務 2　母羊的發情鑑定與輸精 ·· 80
　Task 2　Estrus Identification and Insemination of Ewes ·················· 80

　任務 3　母豬的發情鑑定與輸精 ·· 82
　Task 3　Estrus Identification and Insemination of Sows ·················· 82

　任務 4　雞的人工輸精 ·· 87
　Task 4　Insemination of Hens ··· 87

專案四　胚胎移植 ………………………………………………………… 90
Project IV　Embryo Transfer ……………………………………………… 90

任務 1　羊的胚胎移植 ………………………………………………… 100
Task 1　Embryo Transfer in Sheep …………………………………… 100
任務 2　牛的非手術法胚胎移植 ……………………………………… 103
Task 2　Non-surgical Embryo Transfer in Cattle …………………… 103

專案五　妊娠診斷 ………………………………………………………… 120
Project V　Pregnancy Diagnosis …………………………………………… 120

任務 1　牛的妊娠診斷 ………………………………………………… 135
Task 1　Pregnancy Diagnosis of Cattle ……………………………… 135
任務 2　羊的妊娠診斷 ………………………………………………… 145
Task 2　Pregnancy Diagnosis of Sheep ……………………………… 145
任務 3　豬的妊娠診斷 ………………………………………………… 148
Task 3　Pregnancy Diagnosis of Pigs ………………………………… 148

專案六　接產與助產 ……………………………………………………… 152
Project VI　Delivery and Midwifery ……………………………………… 152

任務 1　牛的接產與助產 ……………………………………………… 161
Task 1　Delivery and Midwifery of Cattle …………………………… 161
任務 2　羊的接產與助產 ……………………………………………… 168
Task 2　Delivery and Midwifery of Sheep …………………………… 168
任務 3　豬的接產與助產 ……………………………………………… 173
Task 3　Delivery and Midwifery of Pigs ……………………………… 173

專案七　繁殖力評價 ……………………………………………………… 177
Project VII　Evaluation of Fecundity ……………………………………… 177

任務　繁殖力評價 ……………………………………………………… 181
Task　Evaluation of Fecundity ………………………………………… 181

參考文獻/References ……………………………………………………… 185

專案一　採　　精
Project Ⅰ　Semen Collection

◎ 專案導學

採精是人工授精技術的首要環節，認真做好採精準備，掌握正確的採精步驟，合理安排採精頻率，才能獲得量多質優的精液，從而提高種公畜（禽）的利用效率。採精的方法很多，如假陰道法（牛、羊和兔等）、手握法（豬）、按摩法（禽類）和電刺激法等。

◎ Project Guidance

Semen collection is the primary component of artificial insemination. In order to improve the utilization efficiency of male animals (poultry) and obtain high-quality semen, we must seriously prepare, correctly grasp the steps and reasonably arrange the frequency of semen collection. There are many methods for semen collection, such as the artificial vagina method (cattle, sheep, rabbits, etc.), the hand holding method (pigs), the massage method (poultry) and the electrical stimulation method.

◎ 學習目標

>>> 知識目標

• 了解雄性動物生殖器官的組成及組織結構特點。
• 掌握雄性動物生殖器官的主要生理功能。
• 理解精子的發生過程。

>>> 技能目標

• 熟練掌握假陰道的安裝與調試。
• 規範採集公牛、公羊的精液。
• 熟練採集公豬的精液。
• 順利完成公禽的採精。

◎ Learning Objectives

>>> Knowledge Objectives

• To understand the composition and structural characteristics of male reproductive organs.
• To master the main physiological functions of male reproductive organs.
• To understand the process of spermatogenesis.

>>> Skill Objectives

• To master the installation and adjustment of artificial vagina.
• To collect sperm of bulls and rams using standard methods.
• To collect sperm of boars skillfully.
• To collect sperm of poultry successfully.

相關知識

一、雄性動物生殖器官的結構與功能

雄性動物的生殖器官由睪丸、附睪、陰囊、輸精管、副性腺、尿生殖道、陰莖和包皮組成。各種雄性動物的生殖器官見圖 1-1。

Relevant Knowledge

1 Structure and functions of male reproductive organs

The male reproductive organs consist of the testis, epididymis, scrotum, duct deferens, accessory sexual glands, urethra, penis and prepuce (Figure 1-1).

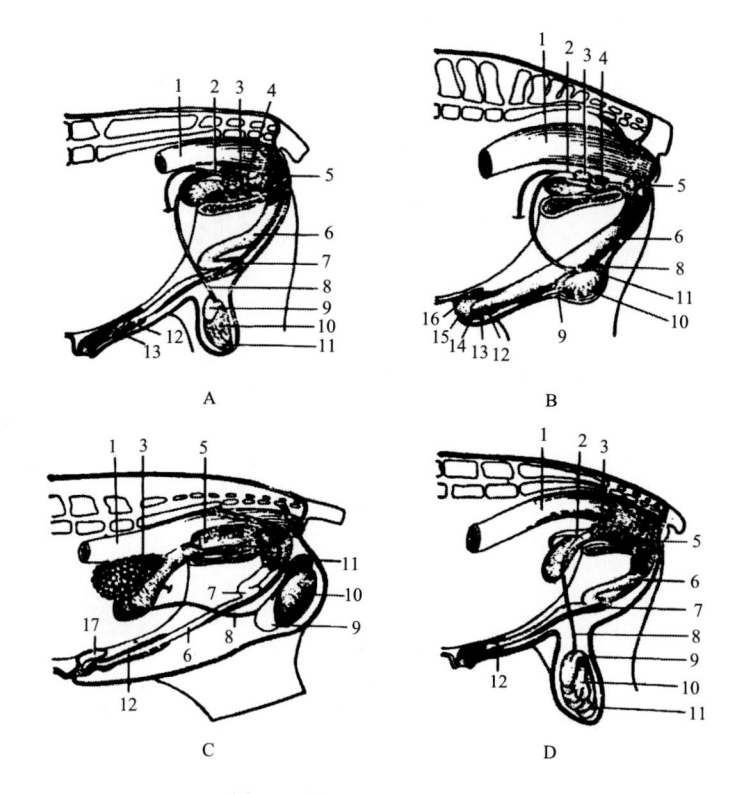

圖 1-1 雄性動物生殖器官示意
Figure 1-1 Reproductive organs of male animals
A. 牛 B. 馬 C. 豬 D. 羊
A. bull B. stallions C. boar D. ram

1. 直腸 2. 輸精管壺腹 3. 精囊腺 4. 攝護腺 5. 尿道球腺 6. 陰莖 7. S狀彎曲 8. 輸精管 9. 附睪頭 10. 睪丸 11. 附睪尾 12. 陰莖游離端 13. 內包皮鞘 14. 外包皮鞘 15. 龜頭 16. 尿道突起 17. 包皮憩室

1. rectum 2. ampulla of duct deferens 3. seminal vesicle gland 4. prostate gland 5. bulbourethral gland 6. penis 7. S-shaped bending 8. duct deferens 9. epididymal head 10. testis 11. epididymal tail 12. penis free end 13. internal prepuce sheath 14. outside prepuce sheath 15. glans penis 16. urethra protrusions 17. prepuce diverticulum

(一) 睪丸

1. 形態和位置

睪丸為雄性動物的性腺，成對存在，呈卵圓形或長卵圓形。不同種屬的雄性動物睪丸的大小和重量有較大差別，豬的睪丸相對較大，為 900～1 000g，占體重的 0.34%～0.38%；牛的相對較小，為 550～650g，占體重的 0.08%～0.09%。禽類睪丸大小和色澤因品種、年齡、生殖季節而有很大變化。雛雞的睪丸只有米粒大小，呈淡黃色或黃色；成年公雞在春季性機能特別旺盛，精子大量形成，睪丸顏色變白，體積變大；當性機能減退時體積變小。

睪丸原位於腹腔內、腎兩側，在胎兒期，由腹腔下降入陰囊。成年公畜如果一側或兩側睪丸並未下降進入陰囊，稱為隱睪，其內分泌機能不受損害，但精子生成會出現異常。公禽的睪丸位於體腔內，形似蠶豆。

2. 組織構造

睪丸表面被覆漿膜，其內為緻密結締組織構成的白膜。白膜由睪丸頭端形成一條結締組織索伸向睪丸實質，構成睪丸縱隔，將睪丸實質分成許多小葉。睪丸小葉由精曲小管盤曲構成，在各小葉頂端匯合成精直小管，穿入縱隔結締組織內形成睪丸網（圖 1-2）。

1.1 Testis

1.1.1 Morphology and location

The testis is the gonad of male animals. It always exists in pairs and the shape is oval or oblong oval. There are many differences in the size and weight of testis among different male animals. The testis of boars are relatively large, which is 900-1 000 g, accounting for 0.34%-0.38% of the body weight. The testis of bulls are relatively small, which is 550-650 g, accounting for 0.08%-0.09% of the body weight. The size and color of poultry testis vary greatly among species, age and reproductive season. The chicken's testis is only a grain size, with a light yellow or yellow colour. In spring, the sexual function of adult cocks is vigorous, with a large number of sperm formed, and the testis become white and large. When the sexual function is going down, the testis becomes smaller.

The testis are originally located in the abdominal cavity, on both sides of the kidneys. During the fetal period, it descends from the abdominal cavity into scrotum. It is called cryptorchidism if one or both testis of adult male animals do not descend into the scrotum. The endocrine function of testis is not affected, but spermatogenesis will be abnormal. The testis of poultry is located in the abdominal cavity and looks like broad beans.

1.1.2 Tissue structure

The surface of testis is covered with serosa, which is a white membrane formed by dense connective tissue. The tunica albuginea forms a connective tissue from the head of the testis to the testicular parenchyma, which constructs the testicu-lar mediastinum and divides the testicular parenchyma into many lobules. The testicular lobules are composed of seminiferous tubules, which converge at the top of each lobule and penetrate into the connective tissue of the mediastinum to form the rete testis (Figure 1-2).

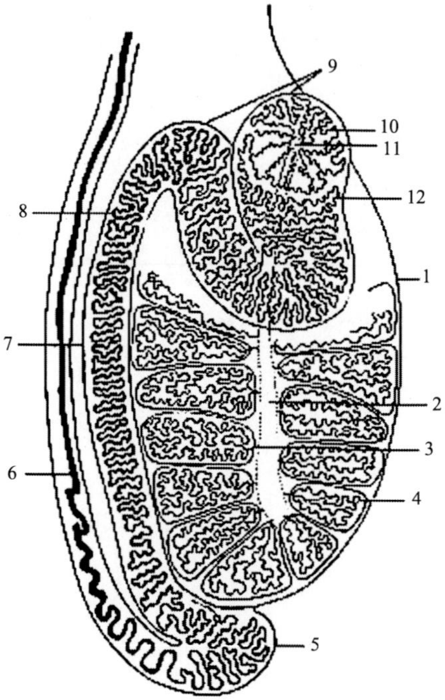

圖 1-2　睪丸及附睪的組織構造
Figure 1-2　Tissue structure of testis and epididymis
1. 睪丸　2. 縱隔　3. 精曲小管　4. 小葉　5. 附睪尾　6. 輸精管　7. 附睪體　8. 附睪管　9. 附睪頭
10. 精直小管　11. 輸出管　12. 睪丸網
1. testis　2. mediastinum　3. seminiferous tubules　4. foliole　5. epididymal tail　6. duct deferens
7. epididymal body　8. epididymal duct　9. epididymal head　10. tubulus rectus　11. duct efferent　12. rete testis

3. 機能

（1）產生精子。精曲小管生殖上皮的生精細胞（精原細胞）經增殖、分裂，最後形成精子。公牛每克睪丸組織平均每天可產生精子1 300萬～1 900萬個，公豬2 400萬～3 100萬個，公羊2 400萬～2 700萬個，公馬1 930萬～2 200萬個。

（2）分泌雄激素。睪丸的間質細胞能分泌雄激素，可激發雄性動物的性慾及性行為，刺激第二性徵，促進生殖器官的發育，維持精子發生及附睪內精子的存活。

1.1.3　Function

(1) Producing sperm. The spermatocytes in the germinal epithelium of seminiferous tubule proliferate, divide, and finally form sperm. Each gram of testicular tissue in a bull can produce 13-19 million sperm per day, 24-31 million in boars, 24-27 million in rams, and 19.3-22 million in stallions.

(2) Androgens secretion. The interstitial cells of the testis can secrete androgens, activating sexual desire and behavior in the male animals, stimulating secondary sexual characteristics, promoting the development of genital organs, maintaining spermatogenesis and sperm survival in epididymis.

(二) 附睾

1. 形態結構

附睾位於睪丸的附著緣，由頭、體、尾三部分組成。附睾頭膨大，由10～30條睪丸輸出管盤曲組成。這些輸出管彙整合一條較粗且彎曲的附睾管，構成附睾體。在睪丸的遠端，附睾體延續並轉為附睾尾，最後逐漸過渡為輸精管。

2. 機能

睪丸是精子儲存和成熟的場所。從睪丸生成的精子，剛進入附睾頭時頸部常有原生質滴，活動微弱，沒有受精能力或受精能力很低。在精子通過附睾的過程中，原生質滴向尾部末端移行並脫落，精子逐漸成熟。由於附睾內的弱酸性（pH為6.2～6.8）、高滲透壓、較低溫度及厭氧的內環境，精子處於休眠狀態。同時附睾管分泌物可提供給精子營養，因此精子在附睾內可儲存較長時間。

(三) 輸精管

輸精管由附睾管延續而來，它與通往睪丸的神經、血管、淋巴管、睪丸提肌共同組成精索。輸精管壁具有發達的平滑肌纖維，射精時憑藉其強而有力的收縮作用將精子排出。

(四) 副性腺

1. 形態結構

副性腺包括精囊腺、攝護腺和尿道球腺（圖1-3）。公禽沒有副性腺。精囊腺成對存在，位於輸精管末端的外側。攝護腺位於精囊腺的後方，由體部和擴散部兩部分組成。尿道球腺成對存在，位於尿生殖道骨盆部後端。

1.2 Epididymis

1.2.1 Morphology and structure

The epididymis is located at the attachment edge of the testis and consists of three parts: head, body and tail. The epididymal head is enlarged and consists of 10–30 testicular efferent tubes coiled. These tubes condense into a thick and curved epididymal tube, forming the epididymal body. At the far end of the testis, the epididymal body continues and turns into the epididymal tail, and eventually transits to duct deferens.

1.2.2 Function

Testis is the place where sperm are stored and matured. Sperms produced from testis often have protoplasmic droplets in the neck when they enter the epididymal head, with weak motility and have no or low fertility. The protoplasmic droplets fall off when the sperms pass through the epididymis, and sperms mature gradually. Because of the weak acidity (pH 6.2-6.8), high osmotic pressure, low temperature and anaerobic internal environment in the epididymis, the sperms are dormant. At the same time, epididymal duct secretions can provide nutrition for sperm, so sperm can be stored in epididymis for a long time.

1.3 Duct deferens

The duct deferens is a continuation of the epididymal duct, which together with the nerves, blood vessels, lymphatic vessels and the testicular muscles form the spermatic cord. The wall of the duct deferens has well-developed smooth muscle fibers, which expel sperm by virtue of its powerful contraction during ejaculation.

1.4 Accessory sexual glands

1.4.1 Morphology and structure

Accessory sexual glands include seminal vesicle gland, prostate gland and bulbourethral gland (Figure 1-3). There is no accessory sexual glands in the poultry. A pair of seminal vesicles glands are located at the outer side of the end of duct deferens. The prostate gland is located behind seminal vesicle gland and consists with body and diffused part. A pair of bulboure-

2. 機能

公畜在射精時，副性腺分泌物與輸精管分泌物混合形成精清，有稀釋精子、沖洗尿生殖道、活化精子、為精子提供營養等作用。

thral gland locates at the back end of the pelvis in the urogenital tract.

1.4.2 Function

When ejaculating, the secretion of accessory sexual glands and duct deferens is mixed to form seminal plasma, which has the functions of diluting sperm, rinsing urogenital tract, activating sperm, providing nutrition for sperm.

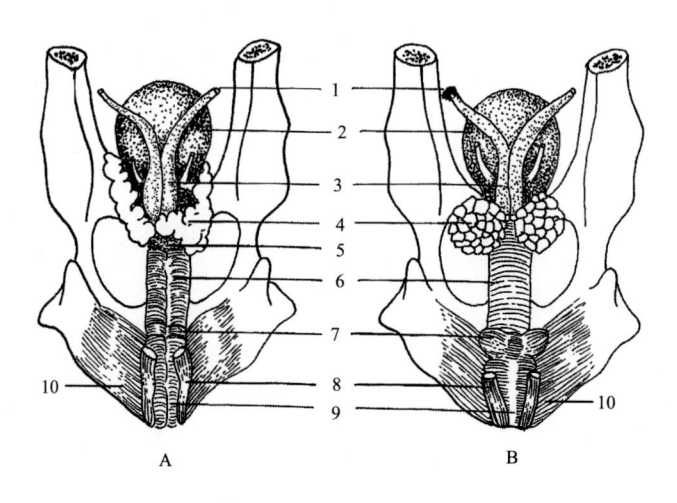

圖 1-3　雄性動物的副性腺

Figure 1-3　Accessory sexual glands of male animals

A. 公牛　B. 公羊

A. bull　B. ram

1. 輸精管　2. 膀胱　3. 壺腹　4. 精囊腺　5. 攝護腺　6. 尿生殖道骨盆部
7. 尿道球腺　8. 坐骨海綿體肌　9. 陰莖縮肌　10. 球海綿體肌

1. duct deferens　2. bladder　3. ampullae　4. seminal vesicle gland　5. prostate gland
6. urogenital tract pelvis　7. bulbourethral gland　8. sciatic sponge muscle　9. penis muscle　10. bulbocavernosus muscle

（五）尿生殖道

尿生殖道為尿液和精液的共同通道，起源於膀胱，終於龜頭，由骨盆部和陰莖部組成（圖1-4）。精阜主要由海綿體組織構成，在射精時可以膨大，關閉膀胱頸，阻止精液流入膀胱，同時阻止尿液混入精液。

1.5 Canalis urogenitalis

The canalis urogenitalis is a common channel for urine and semen. It originates from the bladder and ends in the glans, consists of the pelvis and penis (Figure 1-4).The verumontanum is mainly composed by the sponge tissue, which can be enlarged during ejaculation, close the bladder neck, prevent the semen from flowing into the bladder, and prevent urine from mixing into semen.

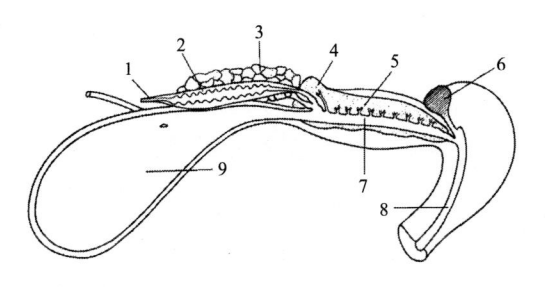

圖 1-4　尿生殖道結構示意

Figure 1-4　Structure of urogenital tract

1. 輸精管　2. 輸精管壺腹　3. 精囊腺　4. 攝護腺體部　5. 攝護腺擴散部
6. 尿道球腺　7. 尿生殖道　8. 陰莖部　9. 膀胱

1. duct deferens　2. ampullae of duct deferens　3. seminal vesicle gland　4. body of prostate gland
5. diffuse part of prostate gland　6. bulbourethral gland　7. urogenital tract　8. penis　9. bladder

(六) 陰莖和包皮

陰莖為雄性動物的交配器官。陰莖由陰莖海綿體和尿生殖道陰莖部組成，分為陰莖根、陰莖體和陰莖頭。陰莖平時柔軟，隱藏於包皮內。交配時勃起，伸長並變得粗硬。

二、 精子的發生

精子的發生是指精子在睪丸內產生的全過程，主要歷經以下三個階段（圖 1-5）。

1. 精原細胞的分裂增殖及精母細胞的形成

這一階段歷時 15～17d。其主要特點為：精原細胞的首次分裂同時產生一個活動的精原細胞和另一個暫時休眠的幹細胞，所有的細胞分裂都是有絲分裂。理論上，一個 A_1 型精原細胞經幾次分裂可生成 16 個初級精母細胞。

2. 精母細胞發育

初級精母細胞形成後，便進入靜止期，期間細胞進行 DNA 複製，此後便進入成熟分裂，初級精母細胞分裂為 2

1.6 Penis and praeputium

The penis is the male reproductive organ of copulation. It consists of the corpus cavernosum and urogenital tract penis, which is divided into root, body and head of the penis. The penis is usually soft and protected in the foreskin. The penis elongates and becomes hard and fully erect in copulation.

2　Spermatogenesis

Spermatogenesis is the whole process of sperm production in testis, which mainly goes through three stages (Figure 1-5).

2.1 The proliferation of spermatogonia and the formation of spermatocytes

This stage lasts for 15-17 days and the main features are as follows. The first division of spermatogonia produces an active spermatocyte and another temporarily dormant stem cell. All cell divisions are mitosis. Theoretically, an A_1 spermatogonium can produce 16 primary spermatocytes after several divisions.

2.2 Development of spermatocytes

After the formation of primary spermatocytes, they enter a static phase which the cells undergo DNA replication and then enter mature division. A primary spermatocyte divides into two secondary spermatocytes,

個次級精母細胞，染色體數目減半。此階段歷時 15～16d。次級精母細胞存在的時間很短，在一天之內分裂成 2 個精子細胞。

3. 精子的形成

精子細胞不再分裂而是經過複雜的形態結構變化形成精子。最初的精子細胞為圓形，以後逐漸變長，發生形態上的急遽變化，最後成為蝌蚪樣的精子。此階段歷時 10～15d。

在精子形成的過程中，由一個 A_1 型活動的精原細胞經過增殖、生長、成熟分裂及變形等階段，最後形成精子這一過程所需的時間，稱為精子發生週期。不同動物的精子發生週期不同，豬為 44～45d，牛為 54～60d，綿羊為 49～50d，山羊為 60d 左右，馬為 49～50d。

and the number of chromosomes is halved. This stage lasts for 15-16 days. The presence of secondary spermatocytes is very short, and it splits into two sperm cells within one day.

2.3　Formation of sperm

Sperm cells no longer divide but evolve into sperm through complex morphological and structural changes. Initially, the sperm cells are round, and then gradually become longer, and the morphology changes dramatically. Finally, they become「tadpole」sperm. This stage lasts for 10-15days.

The time required for spermatogenesis is called spermatogenesis cycle, which goes through the proliferation, growth, mature division and deformation of an A_1 active spermatocytes. The spermatogenesis cycle is 44-45 days for boar, 49-50 days for ram, about 60 days for goat, 49-50 days for stallion.

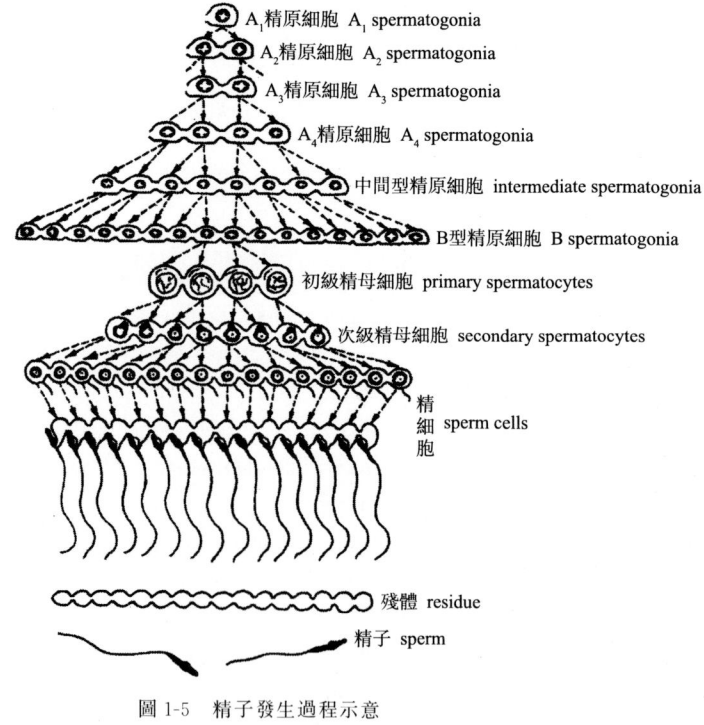

圖 1-5　精子發生過程示意

Figure 1-5　Spermatogenesis

Project I Semen Collection

任 務　採　精
Task　Semen Collection

任務描述 / Task Description

科學合理的採精技術是鮮精產量增加及品質提高的重要環節，有利於保持種公畜（禽）神經興奮的持續性和連貫性。生產中，要根據種公畜（禽）的年齡、季節、體況等因素，嚴格規範採精操作，合理安排採精頻率。

Scientific and reasonable technology of semen collection is an important link in increasing the yield and quality of fresh semen, which is conducive for maintaining the continuity and consistency of the nerve excitement of male animals. During production, according to factors such as the age, season and body condition of the breeding stock (poultry), the semen collection operation should be strictly regulated and the semen collection frequency should be arranged reasonably.

任務實施 / Task Implementation

一、準備工作 / 1　Preparation

（一）採精場地 / 1.1　Site of semen collection

採精場地應固定、寬敞、平坦、安靜、清潔，設有假臺畜或採精架（圖1-6）。理想的採精場地應設有室內和室外兩部分，並與精液品質檢查室、輸精操作室相連或距離很近。

It should be fixed, spacious, flat, quiet and clean, with a dummy or frame for semen collection (Figure 1-6). The ideal site of semen collection should be divided into indoor and outdoor parts, and connected or close to the semen quality inspection room and insemination room.

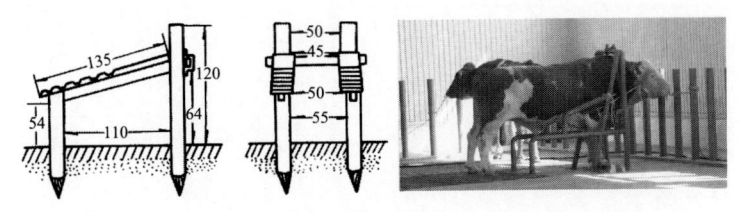

圖1-6　牛的採精架（單位：cm）
Figure 1-6　Mount stall of bull (unit: cm)

（二）臺畜 / 1.2　Dummy

臺畜有真臺畜和假臺畜兩種。公

There are two kinds of dummy, estrous animals

豬採精多採用假臺豬（圖 1-7），採精前對其徹底消毒。公牛多採用真臺牛，有利於刺激公牛的性反射，也可採用假臺牛（圖 1-8），要求大小適宜、堅實牢固。

and fake dummy. Dummy sow is often used in semen collection(Figure 1-7). Estrous cow is usually used to stimulate the sexuality of bulls, but also can use fake dummy cow(Figure 1-8), which is required to be appropriately sized, solid and firm.

圖 1-7 假臺豬

Figure 1-7 Fake dummy sow

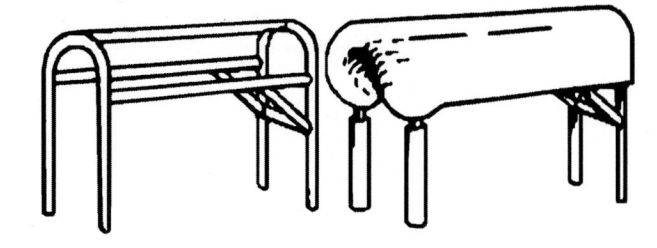

圖 1-8 假臺牛

Figure 1-8 Fake dummy cow

(三) 器材

採精用的所有器材，均應力求清潔無菌，在使用之前要嚴格消毒，使用之後必須洗刷乾淨。

1. 假陰道（牛、羊）

（1）假陰道的結構。假陰道是模擬發情母畜陰道內環境而仿製的人工陰道，主要由外殼、內胎、集精杯、活塞等部件構成（圖 1-9、圖 1-10）。外殼多為硬橡膠或塑膠製成，內胎為彈力強、柔軟的乳膠或橡膠製成，集精杯（瓶）一般用棕色玻璃製成。

1.3 Equipment

All equipment used for sperm collection should be clean and sterile, strictly disinfected before use and cleaned after use.

1.3.1 Artificial vagina(bull, ram)

(1) Structure of artificial vagina

The artificial vagina is made by simulation the environment of estrous female's vagina. It consists of the shell, inner tube, semen collection cup, piston and so on (Figure 1-9, Figure 1-10). Its shell is mostly made of hard rubber or plastic, and the inner tube is made of elastic and soft latex or rubber. The semen collection cup is usually made of brown glass.

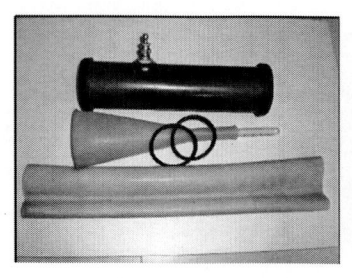

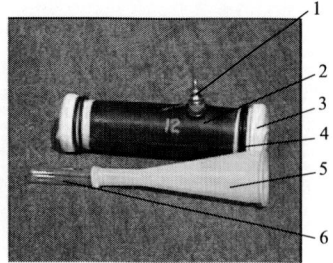

圖 1-9 牛的假陰道結構

Figure 1-9 Structure of bull artificial vagina

1. 活塞 2. 外殼 3. 內胎 4. 膠圈 5. 橡膠漏斗 6. 集精杯

1. piston 2. shell 3. inner tube 4. rubber ring 5. rubber funnel 6. semen collection cup

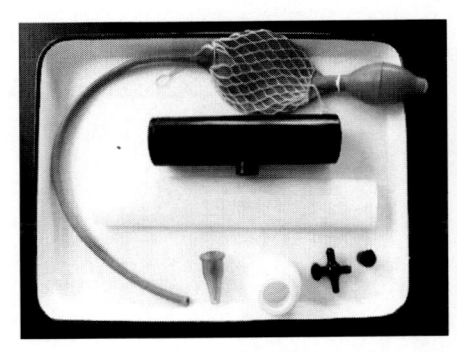

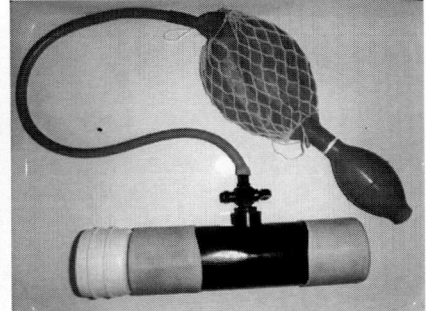

專案一 採 精
Project I Semen Collection

圖 1-10　羊的假陰道結構
Figure 1-10 Structure of ram artificial vagina

（2）假陰道的安裝。

①檢查。安裝前，要仔細檢查外殼是否有裂口、沙眼等，內胎是否漏氣、有無破損，活塞是否完好或漏氣、扭動是否靈活，集精杯是否破裂等。

②安裝內胎。將內胎兩端翻捲於外殼上，要求鬆緊適度、不扭曲，內胎中軸與外殼中軸重合，即「同心圓」，再用膠圈加以固定。安裝好內胎，充氣調試呈 Y 形（圖 1-11）。

③消毒。用長柄鉗子夾取酒精棉球對集精杯消毒，同時由裡向外螺旋式對內胎進行擦拭消毒。採精前，最好用生理鹽水或稀釋液沖洗 1～2 次，並安裝集精杯。

④水肉。由水肉孔向外殼內注入 50～55℃ 的溫水，水量為外殼與內胎容積的 1/3～1/2，主要目的是調節溫度和壓力。

⑤塗抹潤滑劑。用消毒玻璃棒蘸取凡士林由外向內在內胎上均勻塗抹，深度為外殼長度的 1/2 左右。

⑥調壓。如注入水後壓力不夠，可通過充氣調壓使假陰道入口處內胎呈 Y 形。

(2) Installation of artificial vagina

① Check. Before installation, it must be checked carefully to make sure that there are no cracks or trachoma of the shell, leaking or damage of inner tube, flaws or leaking or jam of the piston, crevices of the semen collection cup.

② Installation of inner tube. Both ends of the inner tube are rolled over the shell, making sure that the tension is moderate and there is no distortion. The inner tube axis coincides with the shell axis, and is fixed with a rubber ring. After installing the inner tube, it need to inflate to make a Y-shape(Figure 1-11).

③ Disinfection. Alcohol cotton ball is used to sterilize the inner tube and semen collection cup. Before collecting semen, it is best to rinse with saline or dilute solution 1-2 times, and then install the sterilized semen collection cup.

④ Water injection. The shell is injected with warm water at 50-55℃ by injection hole. The amount of water is about 1/3-1/2 of the volume of the shell and inner tube. The main purpose is to adjust the temperature and pressure.

⑤ Smearing lubricant. Vaseline is used to smear the inner tube from the outside to the depth with a sterile glass rod, about 1/2 of the length of the shell.

⑥ Pressure adjustment. If the pressure is not high enough after water injection, it can be made into Y-shape by air at the entrance of the inner tube.

⑦測溫。用消毒的溫度計插入假陰道內腔，待溫度不變時讀數，一般以38～40℃為宜。

⑦ Temperature measurement. Inserting a sterilized thermometer into the cavity and reading when the temperature is constant, it is generally required to be at 38-40℃.

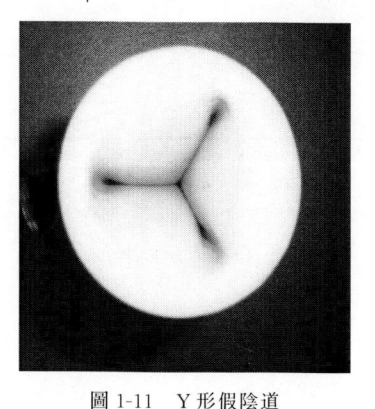

圖 1-11　Y 形假陰道

Figure 1-11　Y-shape artificial vagina

2. 集精杯

（1）豬。將食品保鮮袋或聚乙烯袋放入採精用的保溫杯中，將袋口打開，環套在保溫杯口邊緣，並將精液過濾紙罩在杯口上，用橡皮筋套住（圖 1-12），蓋上蓋子，放入 37℃恆溫箱中預熱。

（2）家禽。集精杯（圖 1-13）必須經高壓消毒後備用。集精瓶內水溫應保持在 30～35℃。

1.3.2　Semen collection cup

(1) Boar

Put the food preservation bag or polyethylene bag into the thermos, open the bag opening to wrap around the edge of the thermos, cover the cup rim with semen filter paper and wrap it with a rubber band (Figure 1-12), cover the lid, then preheat at a 37℃ thermostat.

(2) Poultry

The semen collection cup (Figure 1-13) must be stored after high pressure disinfection. The water temperature in the cup should be maintained at 30-35℃.

圖 1-12　豬用集精杯

Figure 1-12　Semen collection cup for boar

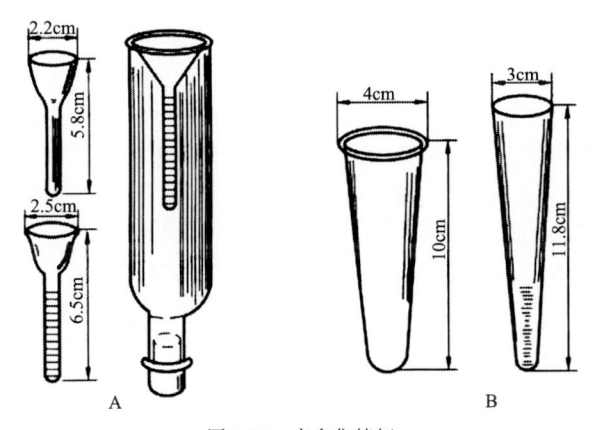

圖 1-13　家禽集精杯
Figure 1-13　Semen collection cup for poultry
A. 雞用集精杯　B. 鴨、鵝用集精杯
A. semen collection cup for cock　B. semen collection cup for drake and gander

（四）種公畜（禽）的準備

採精前，用 0.1％高錳酸鉀溶液清洗公畜包皮部並擦乾。公豬要擠出包皮積尿。公雞在採精前 3～4h 斷水、斷料，並剪去泄殖腔周圍的羽毛。

（五）採精員的準備

採精員應技術熟練，動作敏捷，熟悉公畜的射精特點，並注意人畜安全。指甲應剪短磨光，手臂要清洗消毒等。公豬採精時，採精員要佩戴雙層手套，以減少精液汙染和預防人畜共患病。

二、採精操作

雄性動物的採精方法有很多，主要有假陰道法、手握法、按摩法和電刺激法等。

無論採用哪種方法採精，都要遵循以下原則：

安全：採精過程要確保操作人

1.4　Preparation of male animals

Before semen collection, the prepuce should be cleaned with 0.1% potassium permanganate solution and dried up afterwards. For boars, the accumulate urine should be squeezed out. The rooster must not be fed with any water or feed 3-4 hours before semen collection, and the feathers around the cloaca should be cut off.

1.5　Preparation of technician

The technicians should be skilled, agile, familiar with the characteristics of male ejaculation, and pay attention to the safety of human and animals. The nails should be cut short and polished, the arms should be cleaned and disinfected. In order to reduce semen contamination and prevent zoonosis, the technician must wear double-layer gloves when collecting boar semen.

2　Semen collection

There are many methods to collect sperm of male animals, such as artificial vagina method, hand holding method, massage method and electric stimulation method.

Whatever method used to collect semen, it must follow the following rules:

Security: when collecting semen, it must ensure

員、公畜等的安全,防止陰莖損傷,避免因不良刺激造成性慾下降等問題。

衛生:採精過程精液最容易受到外界因素的影響,因此,採精過程中必須小心操作,保證精液不會受到汙染。

全份:必須收到全份的精液,避免精液損失。

簡便:操作過程要力求簡單,並且容易拆卸、清洗、消毒。

(一) 公牛的採精

公牛的採精多採用假陰道法。採精時,採精員站在臺牛的右側斜後方,當公牛爬上臺牛時,迅速跨前一步,左手迅速拖住包皮,右手持假陰道並調整角度使之與公牛陰莖的伸出方向呈一直線,使陰莖自然插入假陰道內(圖1-14)。當公牛後肢跳起,臀部用力向前一衝,即已射精。

射精後將集精杯向下傾斜,使精液順利流入集精杯。待陰莖自然脫離後立即豎立假陰道,打開氣門,放掉空氣,以充分收集滯留在假陰道內壁上的精液,然後小心取下集精杯,迅速轉移至精液處理室。

the safety of operators and animals to prevent penis damage and avoid the problem of declining sexual desire due to adverse stimuli.

Health: when collecting the semen, the semen is most easily affected by external factors. Therefore, the semen collection must be handled carefully to ensure that the semen is not contaminated.

Full share: it must receive the entire amount of semen to avoid the loss of semen.

Easy: the operation process should be as simple as possible, and the equipment can be easily disassembled, cleaned and disinfected.

2.1 Bull semen collection

Artificial vagina is often used to collect bull semen. The technician stands behind the right side of the dummy cow. When the bull climbs the dummy cow, he steps forward quickly and pulls the foreskin with his left hand, holds the artificial vagina with his right hand and adjusts the angle to make it in line with the direction of bull penis. So that the penis naturally inserts into the artificial vagina(Figure 1-14). When the hind limbs of the bull jump up and the buttocks rush forward forcefully, it has ejaculated.

After ejaculation, the technician tilts the cup downward to make the semen flow into the cup smoothly. After the natural detachment of the penis, he erects the artificial vagina, opens the valve and releases the air to collect the semen that remained on the inside wall. Then the cup should be carefully removed and quickly transferred to the treatment room.

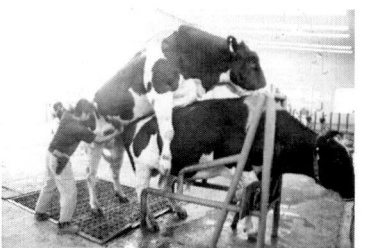

圖 1-14 公牛的採精

Figure 1-14 Bull semen collection

專案一 採精
Project I   Semen Collection

（二）公羊的採精

羊的採精方法與牛相似，羊從陰莖勃起到射精只有幾秒，所以要求操作人員動作敏捷、準確。

採精時牽引公羊接近母羊，用發情母羊刺激公羊。採精者多站在母羊的右後方，右手持假陰道，並用食指固定集精杯，防止脫落。當公羊爬跨母羊時，迅速將假陰道口對準公羊陰莖，方向保持一致，同時用左手迅速將陰莖牽引入假陰道內，但不要觸及陰莖，以免採精失敗，或導致公羊惡癖。公羊有前衝動作即為射精，射精時將集精杯一端適當向下傾斜，以便精液順利流入集精杯中。公羊射精後，待其從母羊身上退下後取出假陰道並豎立，使集精杯一端向下，放掉空氣，然後取下集精杯，運往精液處理室進行精液品質檢查。

（三）公豬的採精

目前，生產上公豬的採精方法主要有手握法和自動採精系統兩種，前者具有操作簡單、可選擇性接取公豬精液等優點，在國內外養豬業被廣泛應用；後者是一種新型的採精系統，採用仿生原理使公豬採精更接近自然狀態，提高了生產效率和精液品質，適合規模化養豬場推廣應用。

1. 手握法採精（圖 1-15）

當公豬性慾旺盛爬跨假臺豬時，採精員左手持集精杯蹲在公豬的左側，右

2.2   Ram semen collection

The semen collection method of ram is similar to the bull. It takes only a few seconds from penile erection to ejaculation for ram, so the technician must be more agile and accurate.

When collecting semen, the technician leads the ram to the estrus ewe so as to stimulate it. The technician generally stands behind the right side of the ewe, holding the artificial vagina with the right hand, and fixing the collection cup with the index finger to prevent it from falling off. When the ram climbs across the ewe, he should quickly points the artificial vagina at the ram's penis in the same direction. At the same time, he should quickly lead the penis into the artificial vagina with the left hand, but do not touch the penis to prevent the failure of collecting semen or lead to the ram evil. The ram's forward action is ejaculation. When the ram ejaculates, the technician should tilt the collection cup to make one end down properly so that the semen will flow smoothly into the collection cup. After the ram ejaculate, they come down from the ewe. The technician should erect the artificial vagina, drop one end of the collection cup, release the air, and then take the collection cup and send it to the semen treatment room for semen quality inspection.

2.3   Boar semen collection

At present, there are two main methods for semen collection of boars: the hand holding and automatic semen collection system. The advantages of the former method are simpl and selectivity of boar semen. And it is widely used in pig husbandry worldwide. The latter is a new type of semen collection system, which uses bionic principle to make closer to the natural state. This improves the production efficiency and semen quality, and it is suitable for large-scale pig farms.

2.3.1   Hand holding method（Figure 1-15）

When the boar is sexually active and climbs across the fake dummy sow, the technician squats on

手呈錐形的空拳於公豬陰莖伸出的同時，將龜頭導入空拳內，順其向前衝力，將陰莖的 S 狀彎曲盡可能拉直，握緊陰莖龜頭防止其旋轉，待充分伸展後，陰莖將停止前衝，開始射精。剛開始射出的清亮液體部分棄去不要，當射出乳白色濃精液時即可收集。公豬射精時間可持續 5～10min，分 2～4 次射出。當公豬開始環顧四周時，說明公豬射精即將結束。採精後，小心將集精杯上的過濾砂紙及上面的膠原蛋白去掉，用蓋子蓋好集精杯，迅速傳遞到精液處理室進行檢查、處理。

the left side of the boar with a cup in the left hand and a conical-shaped empty fist in the right hand. While the penis extends out, the technician leads the glans into the empty fist. Following its forward momentum, he pulls the penis as straight as possible and holds the glans tightly to prevent twisting. After fully stretching, the penis will stop rushing and begin to ejaculate. Discard the bright liquid part emitted at the beginning and collect milk-white semen. Boar ejaculation lasts for 5-10 minutes at 2-4 times. When the boars begin to look around, the ejaculation is about to end. After semen collection, remove the filter sandpaper with the collagen on it carefully and cover semen collection cup. Then transfer to the treatment room for examination and treatment quickly.

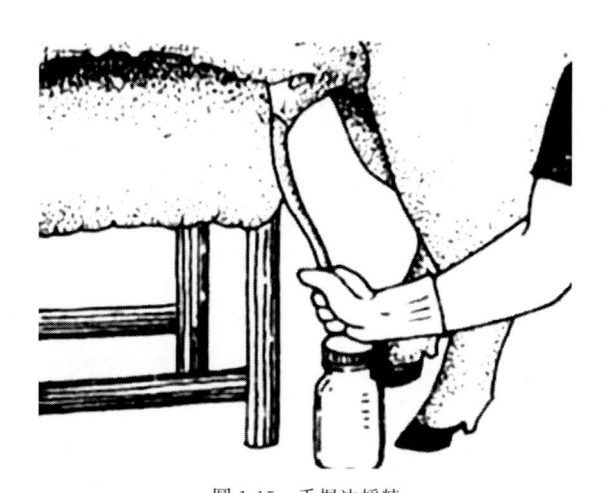

圖 1-15　手握法採精

Figure 1-15　Hand holding method

2. 自動採精系統

自動採精系統主要由採精間、人造仿生假陰道、假臺豬、主控箱組成（圖 1-16）。該採精系統可使公豬在射精時得到切實的生物實體感，延長種公豬的使用年限，避免精液汙染，提高精液品質和生產效率。

2.3.2　Automatic semen collection system

It is mainly composed of the semen collection room, artificial bionic vagina, fake dummy sow and main control box(Figure 1-16). It can make boars get a real sense of biological entity during ejaculation. So it prolongs the service life of boars, avoids semen pollution, improves semen quality and production efficiency.

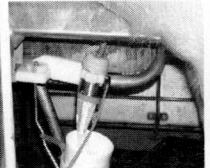

圖 1-16　自動採精系統

Figure 1-16　Automatic semen collection system

（四）公雞的採精

公雞的採精普遍採用背腹式按摩兩人採精法（圖 1-17）。採精時，一人固定公雞，夾於腋下，雙手握住兩腿使其自然分叉，雞頭向後。採精員左手沿公雞背鞍部向尾羽方向撫摩數次，以緩解公雞驚恐並引起性興奮，右手中指和無名指夾集精杯，杯口向外。待公雞有性反射時，左手迅速翻轉，將尾羽向背部壓住，並以拇指與食指跨在泄殖腔上側；右手拇指和食指跨在泄殖腔下側腹部柔軟部，抖動觸摸數次，當泄殖腔外翻露出退化交媾器時，左手拇指與食指立刻輕輕擠壓，公雞就能排精。與此同時，迅速將集精杯口翻向泄殖腔開口處承接精液。採集的精液置於 25～30℃ 的保溫瓶內以備處理。

2.4　Cock semen collection

The back-abdominal massage by two people is widely used in cock semen collection (Figure 1-17). During semen collection, one person clamps the cock under his armpit, and grasps the legs with both hands to make it forked naturally with the cook head backward. The left hand of semen collector strokes several times along the saddle toward the tail feather to relieve its panic and cause sexual excitement. His right hand clamps the cup with the middle finger and ring finger and keep the cup rim outward. When the cock is sexually reflected, the semen collector quickly flips the left hand, presses the tail feather to the back, and cross the upper side of the cloaca with his thumb and index finger. His right thumb and index finger cross the soft part of the abdomen under the cloaca shaking and touching several times. When the cloaca turns out to reveal degraded copulatory organ, the left thumb and index finger immediately squeeze gently, and the cock can ejaculate. At the same time, quickly turn the cup to the cloaca opening to receive semen. Then put the semen in a thermos bottle at 25-30℃.

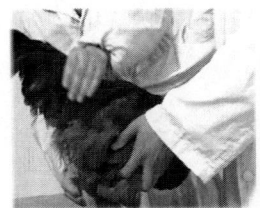

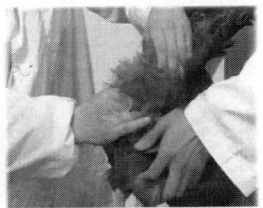

圖 1-17　公雞的採精

Figure 1-17　Cock semen collection

三、採精頻率

公畜（禽）的採精頻率應根據其種類、個體差異、健康狀況、性慾強弱、精子產生數量等確定。生產實踐中，成年公牛每週2～3次；公羊在配種季節內可每天連續採精2～3次，每週5～6d；成年公豬隔天採精1次，青年公豬和老齡公豬以每週採精2次為宜；公雞每週採精4～5次。

3 Frequency of semen collection

The semen collection frequency of male animals should be determined according to their species, individual differences, health status, sexual desire and semen production. In production practice, adult bulls can be collected semen 2-3 times a week. During mating season, the rams can be collected 2-3 times a day, 5-6 days a week. Adult boars are collected once every other day, but young and old boars are collected twice a week. For roosters, collect semen 4-5 times a week.

專案二　精液的處理
Project Ⅱ　Semen Treatment

專案導學

　　精液的品質控制是影響人工授精效果的關鍵技術環節，稀釋、分裝、保存和運輸等環節均可能會受到溫度、細菌、灰塵雜質的汙染而降低精液品質。精液的品質分析又易受環境、檢測人員的技術熟練程度及主觀判斷能力等諸多因素影響而發生偏差，因此精液處理必須嚴格按照規範程序進行。

Project Guidance

　　Quality control of semen is the key links that affect the effect of artificial insemination. Dilution, packaging, preservation and transportation may be affected by temperature, bacteria and dust impurities, which may reduce the quality of semen. The quality analysis of semen is easily affected by many factors, such as environment, technical proficiency and subjective judgment of testers. Therefore, semen treatment must be carried out in strict accordance with the standard procedures.

學習目標

>>> 知識目標

- 了解精液的理化特性。
- 掌握精子的形態結構及生理特性。
- 理解外界因素對精子存活的影響。
- 理解稀釋液的主要成分及作用。
- 理解精液保存的原理。

>>> 技能目標

- 能熟練評定精液的品質。
- 會確定精液的稀釋倍數。
- 能熟練配製稀釋液，會正確稀釋精液。

Learning Objectives

>>> Knowledge Objectives

- To understand the characteristics of semen.
- To master the morphological structure and physiological characteristics of sperm.
- To understand the influence of external factors on sperm survival.
- To understand the main ingredients and functions of semen dilution.
- To understand the principles of semen preservation.

>>> Skill Objectives

- To assess semen quality.
- To calculate the multiple of semen dilution.
- To prepare the diluent and dilute semen correctly.

- 能正確對精液進行保存和運輸。

相關知識

一、精液的組成

精液由精子和精清兩部分組成。精子由睪丸產生，占的比例很小，主要化學成分為核酸、蛋白質和脂類。精清是附睪、副性腺、輸精管壺腹部的分泌物，占精液的80%～90%。精液中90%～98%為水分，乾物質只占2%～10%，在乾物質中約有60%是蛋白質。各種動物精液量的多少，主要取決於副性腺的發達程度。牛、羊的副性腺不發達，精清分泌量少，故射精量就少，但精子密度較大；豬、馬、驢的副性腺比較發達，分泌量多，其射精量也多。

不同動物精清的化學組成差異較大，即使同種動物或同一個體，因採精方法、採精時間及採精頻率等不同，精清成分也有一定的變化。各種動物的精清成分見表2-1。

- To preserve and transport the semen correctly.

Relevant Knowledge

1 Composition of semen

Semen consists of sperm and seminal plasma. Sperm is produced by testis, which accounts for a small proportion of semen. The main chemical constituents of sperm are nucleic acids, proteins and lipids. Seminal plasma is the secretion of epididymis, accessory gonads and vas deferens ampulla, accounting for 80%-90% of semen. 90%-98% of semen is water, and only 2%-10% of semen is dry matter, 60% of dry matter is protein. The amount of semen in animals depends mainly on the development of accessory gonads. The accessory gonads of cattle and sheep are less developed, so the secretion of seminal plasma is less than that of pig and horse. Therefore, the volume of sperm in cattle and sheep is small, but the density of sperm is large. The accessory gonads of pig, horse and donkey are highly developed, thus having more seminal plasma and more semen.

The chemical composition of seminal plasma in different animals is quite different. Even for the same species or the same individual, due to different methods, time and frequency of semen collection, the seminal plasma ingredients also undergo certain changes. The ingredients of seminal plasma in various animals are shown in Table 2-1.

表 2-1　各種動物精清的化學組成
Table 2-1　The components of seminal plasma

成分 Components	動物種類 Animal species			
	牛 Cattle	羊 Sheep/Goat	豬 Pig	雞 Chicken
蛋白質 Protein/(g/dL)	6.8	5.0	3.7	1.2～2.8

(續)

成分 Components	動物種類 Animal species			
	牛 Cattle	羊 Sheep/Goat	豬 Pig	雞 Chicken
果糖 Fructose/ (mg/dL)	460～600	250	9	4
山梨醇 Sorbitol/ (mg/dL)	10～140	26～170	6～8	0～10
肌醇 Inositol/ (mg/dL)	25～46	7～14	380～630	16～20
檸檬酸 Citric acid/ (mg/dL)	620～806	110～260	173	0
甘油磷脂醯膽鹼 Glycerol phosphatidylcholine/ (mg/dL)	100～500	1 100～2 100	110～240	0～40
鈉 Sodium/ (mg/dL)	225±13	178±11	587	352
鉀 Potassium/ (mg/dL)	155±6	89±4	197	61
鈣 Calcium/ (mg/dL)	40±2	6±2	6	10
鎂 Magnesium/ (mg/dL)	8±0.3	6±0.8	5～14	14
氯化物 Chloride/ (mg/dL)	174～320	86	260～430	147

二、精子的形態結構

各種動物精子的形狀、大小及內部結構有所不同，但大體上是相似的（圖2-1）。哺乳動物的精子整體型狀呈蝌蚪狀，主要由頭部、頸部和尾部組成（圖2-2）。精子長度因動物種類不同而有差異，家畜精子的長度為50～90μm。精子的長度和體積與動物自身大小無關，如大鼠精子長約190μm，而大象的精子只有50μm長。

2 Morphological structure of sperm

The shape, size and internal structure of sperm of various animals are different, but they are generally similar(Figure 2-1).The sperm of mammals are tadpole-shaped, consisting of head, neck and tail(Figure 2-2). Sperm length varies according to animal species.Sperm length of livestock is about 50-90μm. The length and volume of sperm have no relevance to the size of the animal. For example, the sperm of the mouse is about 190μm long, while the sperm of the elephant is only 50μm long.

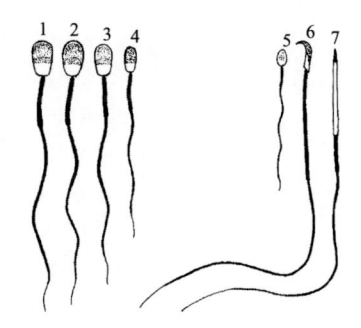

圖 2-1 各種動物精子的形態
Figure 2-1 Morphology of sperms in various animals
1. 牛 2. 豬 3. 羊 4. 馬 5. 人 6. 小鼠 7. 雞
1. bull 2. boar 3. ram 4. horse 5. human 6. mouse 7. rooster

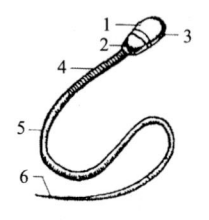

圖 2-2　精子的形態結構

Figure 2-2　Morphological structure of sperm

1. 頭　2. 頸　3. 頂體　4. 中段　5. 主段　6. 末段

1. head　2. neck　3. acrosome　4. middle piece　5. principal piece　6. end piece

1. 頭部

精子的頭部主要由細胞核和頂體構成，核內含有遺傳物質 DNA。核的前端為頂體，是一個由雙層膜組成的帽狀結構，覆蓋在核的前 2/3 部分，靠近質膜的一層稱為頂體外膜，靠近核的一層稱為頂體內膜。頂體內含有多種與受精有關的酶，是一個不穩定的特殊結構，其畸形、缺損或脫落會使精子的受精能力降低或完全喪失。精子頂體異常率高低是評定精液品質好壞的重要指標之一。

2. 頸部

精子的頸部是連接頭部和尾部的部分，呈短圓柱狀。精子頸部長約 $0.5\mu m$，脆弱易斷，特別是在精子成熟過程中或在精液稀釋、保存、運輸時，受到不良影響，極易從頸部斷開，成為頭、尾分離的畸形精子。

3. 尾部

尾部為精子最長的部分，是精子的代謝器官和運動器官。尾部長 $40\sim 50\mu m$，分為中段、主段及末段三部分。精子主要靠尾部鞭索狀波動推動精子向前運動。

2.1　Head

The head of sperm is mainly composed of nucleus and acrosome, and the nucleus contains DNA. The front end of the nucleus is the acrosome, which is a cap-like structure consisting of two layers of membranes covering the first two-thirds of the nucleus. The layer close to the plasma membrane is called the outer acrosome membrane, and the layer close to the nucleus is called the inner acrosome membrane. The acrosome contains a variety of enzymes related to fertilization, which is an unstable structure. The abnormality and defects of acrosome can reduce or completely lose the fertilization ability of sperm. Sperm acrosome abnormality rate is one of the important indicators to evaluate the quality of semen.

2.2　Neck

The neck of sperm is a short cylindrical part connecting the head and tail. The neck is about $0.5\mu m$ long, fragile and easy to break, especially in the process of sperm maturation or in semen dilution, preservation and transportation. The fracture usually leads to a deformed sperm disconnecting from the neck.

2.3　Tail

The tail is the longest part of sperm, which is the metabolic organ and motor organ. The tail length ranges from $40\mu m$ to $50\mu m$, which is divided into three parts: middle piece, principal piece and end piece. Sperm is mainly driven forward by whiplash-like fluctuations of the tail.

三、精子的生理特性

（一）精子的代謝

新陳代謝是精子維持其生命和運動能力的基礎。精子代謝主要是通過糖解和呼吸作用兩種方式進行，這是在不同條件下既有連繫又有區別的代謝過程。

1. 糖解

糖類是精子代謝的主要基質。無論在有氧還是無氧條件下，精子都可以把精清中的果糖或稀釋液中的葡萄糖、乳糖、蔗糖等物質分解成乳酸而釋放出能量，此過程稱為糖解。

2. 呼吸作用

在有氧條件下，精子可將糖解過程中產生的乳酸、丙酮酸等有機酸，通過三羧酸循環徹底分解為 CO_2 和水，產生更多的能量。呼吸旺盛，會使氧和代謝基質消耗過快，造成精子早衰，所以在保存精液時應採取隔絕空氣或充入 CO_2、降低溫度及 pH 等辦法，盡量減少能量消耗，延長其體外存活時間。

（二）精子的運動

運動能力是精子有生命力的重要特徵之一。精子的運動依賴於尾部的擺動。顯微鏡下觀察到的精子運動類型有三種（圖 2-3）：直線前進運動、轉圈運動和原地擺動。其中，直線前進運動是精子正常的運動形式，這樣的精子能運行到受精部位參與受精作用，稱為有效精子。而轉圈運動和原地擺動的精子不具備受精能力。

3 Physiological characteristics of sperm

3.1 Sperm metabolism

Metabolism is the basis for sperm to maintain its life and motor ability. Sperm metabolism is mainly carried out by glycolysis and respiration, which are both related and different metabolic processes under different conditions.

3.1.1 Glycolysis

Carbohydrates are the main substrates of sperm metabolism. Under both aerobic and anaerobic conditions, sperm can decompose fructose in the semen or glucose, lactose and sucrose in dilution into lactic acid to release energy. This process is called glycolysis.

3.1.2 Respiration

Under aerobic conditions, sperm can completely decompose lactic acid, pyruvate and other organic acids produced during glycolysis into carbon dioxide and water through tricarboxylic acid cycle, generating more energy. Vigorous breathing will cause excessive consumption of oxygen and metabolic matrix, resulting in premature sperm death. Therefore, when storing semen, it should be isolated from air or filled with carbon dioxide, with lower temperature and pH, to minimize energy consumption and prolong its survival time in vitro.

3.2 Sperm motility

Motor ability is one of the important indicators of sperm vitality. Sperm movement depends on the tail swing. There are three types of sperm motions observed under a microscope, namely straight forward motion, circle motion, and swing in situ (Figure 2-3). Among them, straight forward motion is the normal form of sperm movement, such sperm can run to the fertilization site to participate in fertilization, called effective sperm. The sperm that moves in circles and swings in situ do not have the ability to fertilize.

图 2-3 精子运动轨迹

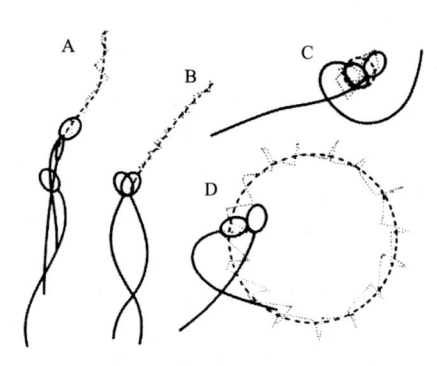

Figure 2-3 Motion trajectory of sperm

A，B. 直线前进运动　C. 原地摆动　D. 转圈运动

A，B. straight forward motion　C. swing in situ　D. circle motion

四、外界因素对精子的影响

影响精子生存的外界因素很多，如温度、光照和辐射、渗透压、pH、电解质浓度、精液的稀释及药品等。在生产中，工作人员要充分了解这些因素的影响，才能在精液的稀释和保存中，控制适宜的条件，延长精子存活和保持受精能力的时间，不断提高人工授精技术的应用效果。

1. 温度

精子对温度十分敏感，体温是保持精子正常代谢和运动的最适温度，哺乳动物的体温为 37～38℃，禽类为 40℃。但在这种温度下，不利于精子的长时间保存。精子对高温的耐受性差，一般不超过 45℃。当温度超过这一限度时，精子经过短促的热僵直后迅速死亡。在低温环境下，精子的代谢和运动受到抑制，能量消耗减少。若精液由体温急遽降至 10℃以下时，精子受到冷打击，不可逆地失去活力

4　Influence of external factors on sperm

Many external factors affect sperm's survival, such as temperature, light and radiation, osmotic pressure, pH, electrolyte concentration, semen dilution and medicine etc. In production process, the staff should fully understand the impact of these factors in order to control the appropriate conditions in the dilution and preservation of semen, prolong the time of sperm survival and fertilization ability, and continuously improve the efficiency of artificial insemination.

4.1　Temperature

Sperm is very sensitive to temperature. Body temperature is the optimum temperature to maintain normal sperm metabolism and movement. The body temperature of mammals is between 37-38℃, and that of poultry is 40℃. However, at this temperature, it is not conducive to the long-term preservation of sperm. The tolerance of sperm to high temperature is poor, usually no higher than 45℃. When the temperature exceeds this limit, the sperm dies quickly after a short thermal stiffness. At low temperatures, sperm metabolism and movement are inhibited, its energy consumption is reduced. If the temperature drops sharply from body

而不能復甦，這種現象稱為冷休克。在超低溫（－196℃）環境中，精子的代謝和運動基本停止，生命處於靜止狀態，可進行長期保存。解凍後，精子活力復甦且不喪失受精能力。

2. 光照和輻射

可見光、紫外光及各種放射性射線均會對精子的活力產生不利影響。直射的日光可提升精子的攝取氧能力，會加速精子的呼吸和運動，從而縮短精子壽命。紫外線對精子的影響取決於它的強度，其中波長 240 nm 的紫外光對精子的危害最大。螢光燈發射的紫外線，對精子也有不利影響。因此，在採精和精液處理時，應盡量減少光的照射，裝精液的容器最好採用棕色瓶。

3. 滲透壓

精子只有與其周圍的液體（如稀釋液）保持等滲，才能維持細胞的完整性和正常的生理活動。周圍液體的滲透壓過高時，精子內的水分向外滲出，精子會失水皺縮，嚴重時會死亡；而在低滲環境中，水分向精子內滲透，使精子膨脹、細胞膜破裂而導致精子死亡。

4. pH

精液 pH 的變化可明顯地影響精子的代謝和活動力，pH 可用 pH 試紙測定。在弱酸性環境中，精子的代謝和活動受到抑制；反之，當 pH 升高時，精子代謝和呼吸增強，運動和能量消耗加劇，精子壽命相對縮短。因此，弱酸性環境更有利於精液保存，可通過向精液

temperature to 10℃, the sperm will be damaged by the cold environment, irreversibly lose vitality and can not recover. This phenomenon is called「cold shock」. In ultralow temperature (－196℃) environment, sperm metabolism and movement are almost stopped, its life is in a static state, and it can be preserved for a long time. After thawing, sperm motility is restored without loss of fertilization capacity.

4.2 Light and radiation

Visible light, ultraviolet light and various radioactive rays can adversely affect the vitality of sperm. Sunlight can improve the oxygen uptake of sperm, accelerate the breathing and movement, and shorten its life. The effect of ultraviolet light on sperm depends on its intensity, and ultraviolet light with a wavelength of 240 nm is the most harmful. Ultraviolet light emitted by fluorescent lamps also has an adverse effect on sperm. Therefore, in the process of semen collection and semen treatment, the irradiation of light should be minimized, and the container for the semen is best to use brown bottles.

4.3 Osmotic pressure

Sperm can only maintain its cell integrity and normal physiological activities when it remains isotonic with its surrounding fluid (such as diluent). When the osmotic pressure of the surrounding liquid is too high, the sperm will lose water and shrink, and will die in severe cases. In the hypotonic environment, the water will penetrate into the sperm, causing the swell and cell membrane rupture, leading sperm to death.

4.4 pH

Changes in semen pH can significantly affect sperm metabolism and motility, and pH can be measured using pH paper. In a weakly acidic environment, sperm metabolism and activity is inhibited. Conversely, when pH rises, sperm metabolism and respiration is enhanced, movement and energy consumption increased, so the sperm life is relatively shortened. Therefore, a weakly acidic environment is more conducive to semen preservation, and the pH can be

中充入飽和 CO_2 氣體或使用碳酸鹽等方法降低 pH。

精子適宜的 pH 範圍因動物種類不同而有差異，一般為：牛 6.9～7.0，羊 7.0～7.2，豬 7.2～7.5，家兔 6.8，雞 7.3。

5. 電解質濃度

精子的代謝和活力受環境離子類型和濃度的影響。一定量的電解質對精子的正常刺激和代謝是必要的，尤其是一些弱鹼性鹽類，如檸檬酸鹽、磷酸鹽等溶液，對維持精液 pH 的相對穩定具有重要作用。但高濃度的電解質易破壞精子與精清的等滲性，易對精子造成損害。

6. 精液的稀釋

精液經過稀釋處理後，精子的代謝和運動加強，受精能力會增強。在進行高倍稀釋時，會使精子膜通透性增大，影響精子的代謝和生存。在稀釋液中加入卵黃並且採用分步稀釋的方式，可減少高倍稀釋對精子的有害影響。稀釋精液時，應該將稀釋液沿著容器壁緩緩加入精液中，邊加入邊搖晃，以防稀釋打擊。

7. 藥品

一些藥品對精子具有保護作用。例如，向精液或稀釋液中加入適量抗生素、磺胺類藥物，能抑制病原微生物的繁殖，有利於精子的保存；在稀釋液中加入適量甘油、二甲基亞碸，可緩解冷凍過程對精子的傷害。但是，某些防腐消毒藥品，如酒精、煤皂酚等對精子有害。所以，在人工授精操作中，工作人

reduced by filling semen with saturated carbon dioxide or using carbonate method etc.

The appropriate pH range of sperm varies depending on the animal species, generally, 6.9-7.0 for bull, 7.0-7.2 for ram, 7.2-7.5 for boar, 6.8 for male rabbit, 7.3 for cock.

4.5 Electrolyte concentration

The metabolism and motility of sperm are affected by the type and concentration of environmental ions. A certain amount of electrolytes is necessary for the normal stimulation and metabolism of sperm, especially some weak alkaline salts, such as citrate, phosphate and other solutions, which play an important role in maintaining the stability of semen pH. However, the high concentration of electrolyte will easily destroy the isotonicity of sperm and seminal plasma, and it is easy to cause damage to sperm.

4.6 Dilution of semen

After dilution, the metabolism and movement of the sperm are strengthened, and its fertilization ability is enhanced. When high-dilution is performed, the permeability of the sperm membrane is increased, which affects sperm metabolism and survival. The harmful effects of high-dilution on sperm can be reduced by adding egg yolk into the dilution solution and adopting a step-by-step dilution. When the semen is diluted, the dilution should be slowly added to the semen along the wall of the container and shaken while adding to reduce the direct impact of dilution.

4.7 Medicine

Some medicines have protective effects on sperm. For example, adding an appropriate amount of antibiotics or sulfonamides to semen or diluent can inhibit the reproduction of pathogenic microorganisms and facilitate sperm preservation, adding appropriate amount of glycerin and dimethyl sulfoxide to diluent can alleviate the damage to sperm in freezing process. However, certain antiseptic and disinfecting drugs, such as alcohol and lysol, are harmful to sperm. Therefore, in

員既要保持所用器械清潔無菌，又要避免將消毒藥液混入精液中。

五、精液的稀釋

在精液中添加一定數量的、適宜於精子存活並保持其受精能力的溶液，稱為精液的稀釋。精液稀釋的目的在於擴大精液容量、提高受配母畜頭數、延長精子的存活時間和受精能力、利於精液的保存與運輸。

(一) 稀釋液的成分及作用

1. 稀釋劑

稀釋劑主要用於擴大精液的容量，要求稀釋劑與精液具有相同的滲透壓。多採用等滲生理鹽水、5％葡萄糖等。

2. 營養劑

營養劑主要是為精子提供營養，補充精子生存和運動所消耗的能量。常用的營養劑有糖類、卵黃、乳類等。

3. 保護劑

（1）緩衝物質。用以保持精液適當的pH。精液在保存過程中，隨著代謝產物（乳酸和CO_2等）的累積，精液的pH會逐漸下降，下降到一定程度時，精子甚至會發生酸中毒，不可逆地失去活力。因此，向精液中添加一定量的緩衝物質，可使pH穩定在一定範圍之內。常用的無機緩衝劑主要有檸檬酸鈉、磷酸二氫鉀、酒石酸鉀鈉等；有機緩衝劑有三羥甲基胺基甲烷（Tris）和乙二胺四乙酸二鈉（EDTA）。

（2）防冷物質。具有防止精子冷休克的作用。在精液保存過程中常需要降

the artificial insemination operation, the operator must keep the instruments used clean and sterile, and avoid the mixing-up of disinfectant liquid and semen.

5 Semen dilution

Adding a certain amount of solution in the semen, which is suitable for sperm survival and maintaining fertilization ability is called semen dilution. The purpose of semen dilution is to expand the semen volume, increase the number of mated females, prolong the survival time and fertilization ability of sperm, and facilitate the preservation and transportation of semen.

5.1 Composition and functions of diluent

5.1.1 Dilutant

It is mainly used for enlarging the volume of semen, requiring the same osmotic pressure between dilutant and semen. Isotopic physiological saline and 5% glucose are often used.

5.1.2 Nutrients

It mainly provides nutrition for sperm and supplements the energy consumed by its survival and movement. Sugar, egg yolk and milk are usually used as nutrients.

5.1.3 Protective agents

5.1.3.1 Buffer substances

Buffer substances are used to maintain the proper pH of semen. In the process of semen preservation, with the accumulation of metabolites (lactic acid and carbon dioxide, etc.), the pH of semen will gradually decrease, and when it drops to a certain extent, the sperm will even be acidosis, which makes the sperm irreversibly lose its vitality. Therefore, adding a certain amount of buffer substances to semen can stabilize the pH within a certain range. The common inorganic buffers are sodium citrate, potassium dihydrogen phosphate, potassium sodium tartrate, etc. The organic buffers are Tris and EDTA.

5.1.3.2 Anti-cold substances

It has the effect of preventing sperm from cold

溫處理，尤其是從 30℃ 急遽下降到 10℃ 時，因精子內含有的縮醛磷脂在低溫下容易凝固，影響精子的正常代謝，造成精子不可逆的冷休克而喪失活力。卵黃和乳類含有卵磷脂，其熔點低，在低溫下不易被凍結，可防止精子冷休克而起到保護作用。

（3）抗凍物質。可以防止冷凍過程中「冰晶化」。精液在冷凍和解凍過程中，精液所經歷的固、液態之間的轉化對精子的危害很大。甘油、二甲基亞碸可有效緩解這種危害，是生產中常用的抗凍保護劑。

（4）抗菌物質。具有抗菌作用。在採精和精液處理過程中，向稀釋液中加入一定量的抗生素可以防止精液遭受微生物的汙染。常用的抗生素有青黴素、鏈黴素、林可黴素、卡那黴素、恩諾沙星。

（5）非電解質和弱電解質。向精液中添加適量的非電解質或弱電解質物質，具有降低精清中電解質濃度的作用。常用的非電解質和弱電解質有各種糖類、胺基乙酸等。

4. 其他添加劑

這類添加劑主要用於改善精子外在環境的理化特性，以及母畜生殖道的生理機能，從而提高受胎率。常用的有酶類、激素類和維他命類物質。如 β-澱粉酶、催產素、前列腺素、維他命 B_1、維他命 B_2、維他命 B_{12}、維他命 C 等。

shock. The cooling process is usually required during semen preservation. Once the temperature drops from 30℃ to 10℃ sharply, the plasmalogen contained in sperm easily condenses at this process, affecting the normal metabolism of sperm, causing the sperm to be irreversibly cold shocked and lose its vitality. Egg yolk and milk contain lecithin, which has a low melting point and is not easily frozen at low temperatures can play a protective role in protecting sperm from cold shock.

5.1.3.3 Antifreeze substances

It is possible to prevent the formation of ice crystallization during the freezing process. During the freezing and thawing process of semen, the conversion between the solid and liquid phases is very harmful to the sperm. Glycerin and dimethyl sulfoxide can effectively alleviate this hazard and are commonly used as antifreeze protectants in production.

5.1.3.4 Antibacterial substances

It has antimicrobial effect. In the process of semen collection and semen treatment, adding a certain amount of antibiotics to the diluent can prevent semen from being contaminated by microorganisms. The commonly used antibiotics are penicillin, streptomycin, lincomycin, kanamycin and enrofloxacin.

5.1.3.5 Non-electrolytes and weak electrolytes

Adding appropriate amount of non-electrolytes or weak electrolyte substances to semen can reduce the concentration of electrolytes in semen. The commonly used non-electrolytes and weak electrolytes are various sugars, aminoacetic acids, etc.

5.1.4 Other additives

These additives are mainly used to improve the physical and chemical properties of the external environment of sperm, as well as the physiological functions of the genital tract of female animal, so as to improve the conception rate. The commonly used additives are enzymes, hormones and vitamins. For example, β-amylase, oxytocin, prostaglandin, vitamin B_1, vitamin B_2, vitamin B_{12}, vitamin C, etc.

（二）稀釋液的種類

1. 現用稀釋液

現用稀釋液適用於精液稀釋後立即輸精用，不進行保存。以單純擴大精液容量、增加輸精母畜頭數為目的。此類稀釋液常以簡單的等滲糖類和乳類溶液為主，也可選用生理鹽水。

2. 常溫保存稀釋液

常溫保存稀釋液適用於精液的常溫短期保存。以糖類和弱酸鹽為主體，一般 pH 控制在 6.35 左右。

3. 低溫保存稀釋液

低溫保存稀釋液適用於精液的低溫保存。以卵黃和乳類為主體，具有抗冷休克的作用。

4. 冷凍保存稀釋液

冷凍保存稀釋液適用於精液冷凍保存。其稀釋液成分較為複雜，除糖類、卵黃外，還應添加甘油或二甲基亞碸抗凍劑。

生產中選用稀釋液時，應根據用途、動物種類及精液的保存時間、保存方法等進行綜合考慮，選擇來源廣、成本低、效果好、易配製的稀釋液配方。

（三）稀釋倍數

精液進行適當的稀釋可以提高精子的存活率，但如果稀釋倍數超過一定限度，其存活率就會受到影響。精液稀釋倍數應根據精液的品質來確定，尤其是精子的活力和密度，還有每次輸精所需要的有效精子數、稀釋液的種類和保存方法等。

（1）牛、羊精子密度大，稀釋倍數可大一點，而豬、馬精子密度小，不宜

5.2 Types of diluents

5.2.1 Simple diluent

It is suitable for insemination immediately after semen dilution and is not used for preservation. Using this type of diluent is to expand the semen volume and increase the number of inseminated female animals. Simple isotonic sugars and dairy solutions are often used in the preparation of such diluents. Normal saline can also be used.

5.2.2 Preservation diluent at room temperature

It is suitable for short-term storage of semen at room temperature. The main components are sugars and weak acid salts, and the pH is generally controlled at about 6.35.

5.2.3 Preservation diluent at low temperature

It is suitable for cryopreservation of semen. Egg yolk and milk are the main components, which have the function of anti-cold shock.

5.2.4 Cryopreservation dilution

It is suitable for cryopreservation of semen. The composition of the diluent is complicated, and in addition to sugars and egg yolk, glycerin or dimethyl sulfoxide antifreeze should be added.

When choosing a diluent in production, comprehensive consideration should be made according to the purpose, animal species, storage time and storage method of semen, etc., and a diluent formula with a wide range of sources, low cost, good effect, and easy preparation should be selected.

5.3 Dilution ratio

Proper dilution of semen can improve the survival rate of sperm, but if the dilution ratio exceeds a certain limit, the survival rate will be affected. The dilution ratio should be determined according to the quality of semen, especially the motility and density of sperm, the number of effective sperm needed for each insemination, as well as the type of diluent and the method of preservation.

(1) Sperm density of bull and ram is large, and it can be highly diluted, while boar and male horse is

作高倍稀釋。

（2）乳類稀釋液可作高倍稀釋，糖類稀釋液不宜作高倍稀釋。

（3）冷凍保存稀釋倍數應低一些，液態保存稀釋倍數可高些。

牛精液耐稀釋的潛力很大，但生產上僅稀釋10～40倍。山羊與牛相似。綿羊精液稀釋後1h，受精率就會有所下降，因此常不作稀釋即用於輸精。公豬精液一般稀釋2～4倍。

六、精液保存的原理

精液保存的目的是延長精子在體外存活的時間，便於長途運輸，從而擴大精液的使用範圍。精液保存的方法，按保存溫度不同可分為常溫（15～25℃）保存、低溫（0～5℃）保存和冷凍（－196℃）保存三種。

1. 常溫保存的原理

常溫保存又稱為室溫保存或變溫保存。常溫保存不需要特殊設備，簡單易行，便於普及和推廣，適用於各種動物精液的短期保存，尤其適合公豬精液的保存。其保存原理是：精子在弱酸性環境中，其活動受到抑制，能量消耗減少，而當pH恢復到中性，精子活力即可復甦。因此，可在稀釋液中加入弱酸性物質（如己酸），把pH調整到6.35左右，從而抑制精子的活動。亦可向稀釋液內充入一定量的CO_2氣體，溶入水形成碳酸，變成弱酸性環境，達到短期保存精液的目的。不同酸類物質對精子產生的抑制區域和保護效果不同，一般認為有機酸好於無機酸。

small in density and not suitable for high dilution.

(2) Milk diluents are suitable for high dilution, while sugar diluents are not suitable for high dilution.

(3) The dilution ratio of cryopreservation should be lower and that of liquid preservation could be higher.

Bull semen has a great potential for dilution resistance, but it is only 10-40 times diluted in production. Goats are the same as cattle in that respect. The fertilization rate of sheep decreases in one hour after diluting, so it is not suitable for dilution. Boar semen is generally diluted 2-4 times.

6 Principle of semen preservation

The purpose of semen preservation is to prolong the survival time of sperm in vitro, facilitate long-distance transportation, and expand the range of application. According to the storage temperature, it can be divided into normal temperature preservation (15-25℃), low-temperature preservation(0-5℃) and cryopreservation(－196℃).

6.1 Principle of normal temperature preservation

The normal temperature preservation is also called room temperature or variable temperature preservation. It is simple and easy to be popularized. It is suitable for short-term preservation of animal semen, especially boar semen. The principle of semen preservation is that sperm activity is inhibited and energy consumption is reduced in weak acid environment, and sperm motility can be recovered once the pH is returned to neutral. Therefore, weak acid substances (such as hexanoic acid)can be added into the diluent to adjust the pH to about 6.35, so as to inhibit sperm activity. Moderate carbon dioxide is filled into the diluent to dissolve into water to form carbonic acid, which will create a weak acid environment, so as to achieve the purpose of short-term preservation of semen. Different acids have different inhibitory regions and protective effects on sperm.

2. 低溫保存的原理

低溫保存是將稀釋後的精液置於 0～5℃ 的環境中保存，保存時間通常比常溫保存時間長，但公豬的精液不如常溫保存效果好。低溫保存原理為：當溫度緩慢降至 0～5℃ 時，精子呈現「休眠」狀態，精子代謝機能和活動力減弱，當溫度回升後，精子又逐漸恢復正常的代謝機能而不喪失其受精能力。

3. 冷凍保存的原理

精液冷凍保存是用液氮（－196℃）或乾冰（－79℃）作冷源，達到長期保存的目的。該方法保存時間長，精液的使用不受時間和地域的限制，是一種比較理想的保存方法，對人工授精技術的推廣及現代畜牧業的發展都具有十分重要的意義。目前，牛的冷凍精液在生產上的普及率已達到 100%。其保存原理為：在超低溫下，精子運動和代謝完全停止，生命以「靜止」狀態保存下來，當溫度回升後，又能復甦且不喪失受精能力。但精子的復甦率只有 50%～70%，部分精子在冷凍過程中死亡。

七、液氮的特性及液氮罐的結構

（一）液氮的特性

液氮是空氣中的氮氣經分離、壓縮形成的一種無色、無味、無毒的透明液體，沸點為－195.8℃，每升液氮為 0.8kg 左右。液氮具有超低溫性，可抑制精子代謝和細菌繁殖，能長期保存精液；液氮具有很強的膨脹性，當溫度達到 15℃ 時，在標準大氣壓下，1L 液氮可氣化為

6.2 Principle of low-temperature preservation

Low-temperature preservation is to store the diluted semen in the environment of 0-5℃. The preservation time is usually longer than that at normal temperature, but effect of boar semen is not as good as that of normal temperature. The principle of low-temperature preservation is as follows: when the temperature drops slowly to 0-5℃, the sperm presents a「dormant」state, and its metabolic function and activity are weakened. When the temperature rises, the sperm gradually returns to normal metabolic function without losing fertilization ability.

6.3 Principle of cryopreservation

Semen cryopreservation is to use liquid nitrogen (－196℃) or dry ice (－79℃) as cold source to achieve the purpose of long-term preservation. It is an ideal method to preserve semen for a long time. It is of great significance to the popularization of artificial insemination technology and the development of modern animal husbandry. At present, the popularization rate of frozen semen in cattle has reached 100%. The principle of cryopreservation is as follows: under ultra-low temperature, sperm movement and metabolism are completely stopped, life is preserved in a「static」state, and when the temperature rises, it can recover without losing fertilization ability. However, the recovery rate of sperm was only 50%-70%, and some sperm died during freezing.

7 Characteristics of liquid nitrogen and structure of liquid nitrogen tank

7.1 Characteristics of liquid nitrogen

Liquid nitrogen is a colorless, odorless and nontoxic transparent liquid formed by the separation and compression of nitrogen in the air. The boiling point is －195.8℃, and the weight of liquid nitrogen is about 0.8 kg per liter.

Liquid nitrogen has ultra-low temperature, which can inhibit sperm metabolism and bacteria reproduction, and can preserve semen for a long time. Liquid

680L 氮氣，膨脹率為 680 倍；液氮具有揮發性，當液氮用量大時，要注意通風，以防窒息。

（二）液氮罐的結構

液氮罐可分為儲存罐和運輸罐兩種。儲存罐主要用於室內液氮的靜置儲存，容量大小不等，大的可達數百升，小的不到 1L，一般的人工授精站適宜用 10～30 L 的中型罐。為了滿足運輸條件，液氮運輸罐增加了專門的防震設計，但也應避免劇烈的碰撞和震動。

液氮罐由外殼、內槽、夾層、頸管、蓋塞、提筒及外套構成（圖 2-4）。

nitrogen has strong expansibility, under standard atmospheric pressure, when the temperature reaches 15℃, one liter liquid nitrogen can be gasified into 680 liters nitrogen. Liquid nitrogen is volatile, when the amount of liquid nitrogen is large, pay attention to ventilation to prevent suffocation.

7.2 Structure of liquid nitrogen tank

There are two types of nitrogen tanks, storage tank and transport tank. The storage tank is mainly used for the static storage of liquid nitrogen indoors. The capacity varies from less than 1 liter to several hundred liters. It is suitable for general artificial insemination station to use 10-30 liters medium-sized tank. In order to meet the transportation condition, the special shock proof design is added to the transport tank, but violent collision and vibration should be avoided.

The liquid nitrogen tank is composed of shell, inner groove, interlayer, neck tube, cap plug, lifting cylinder and coat (Figure 2-4).

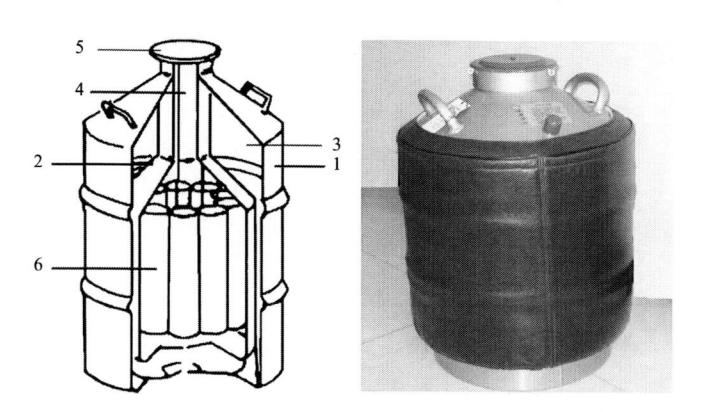

圖 2-4　液氮罐
Figure 2-4　The liquid nitrogen tank
1. 外殼　2. 內槽　3. 夾層　4. 頸管　5. 蓋塞　6. 提筒
1. shell　2. inner groove　3. interlayer　4. neck tube　5. cap plug　6. lifting cylinder

1. 外殼

液氮罐的罐壁由內、外兩層構成，外層稱為外殼，內層稱為內膽，一般由

7.2.1 Shell

The wall of liquid nitrogen tank is composed of inner and outer layers. The outer layer is called the

堅硬的合金製成。

2. 內槽

液氮罐內層中的空間稱為內槽。內槽的底部有底座，供固定提筒用，可將液氮及冷凍精液儲存於內槽中。

3. 夾層

內外兩層間的空隙為夾層。為增加罐的保溫性，夾層被抽成真空，在夾層中裝有絕熱材料和吸附劑（如活性炭），以吸收漏入夾層的空氣，從而增加了罐的絕熱性能。

4. 頸管

頸管以絕熱黏合劑將罐的內外兩層連接，並保持有一定的長度。頂部有罐口，其結構既要有孔隙能排出液氮蒸發出來的氮氣以保證安全，又要有絕熱性能以盡量減少液氮的氣化量。

5. 蓋塞

蓋塞由絕熱性能良好的塑膠製成，以阻止液氮的蒸發，又可固定儲精提筒的手柄。

6. 提筒

提筒是存放凍精的裝置。提筒的手柄由絕熱性能良好的塑膠製成，既能防止熱量向液氮傳導，又能避免取凍精時凍傷。提筒的底部有多個小孔，以便液氮滲入其中。

7. 外套

中、小型液氮罐為了攜帶方便，有

shell, and the inner layer is called the liner. It is generally made of hard alloy.

7.2.2 Inner groove

The space in the inner layer of the liquid nitrogen tank is called the inner groove. There is a base at the bottom of the inner tank for fixing the lifting cylinder, which can store liquid nitrogen and frozen semen in the inner tank.

7.2.3 Interlayer

The gap between the inner and outer layers is the interlayer. In order to improve the thermal insulation of the tank, the interlayer is pumped into a vacuum, the insulation material and adsorbent (such as activated carbon) are installed in the interlayer to absorb the air leaking into the interlayer, thus increasing the thermal-insulation performance of the tank.

7.2.4 Neck tube

The neck tube connects the inner and outer layers of the tank with insulating adhesive and keeps a certain length. There is an opening on the top of the tank. The structure should not only have holes to discharge the nitrogen to ensure safety, but also have insulation performance to reduce the gasification amount of liquid nitrogen as much as possible.

7.2.5 Cap plug

The cap plug is made of plastic with good insulation performance to prevent the evaporation of liquid nitrogen, and can fix the handle of lifting cylinder.

7.2.6 Lifting cylinder

The lifting cylinder is a device for storing frozen semen. The handle of lifting cylinder is made of plastic with good insulation performance, which can not only prevent the heat from transmitting to liquid nitrogen, but also avoid frostbite when taking frozen semen. There are many small holes at the bottom of the lifting cylinder so that liquid nitrogen can penetrate into it.

7.2.7 Coat

For the convenience of carrying, the small and

一外套並附有斜背用的皮帶。

　　液氮罐都應該具有絕熱性能好、堅固耐用、使用方便等特點，這樣才能更好地滿足生產的需要。

（三）液氮罐的使用

1. 使用前要認真檢查

　　新購或長期未用的液氮罐，必須外部無破損、無異常、內部乾燥無異物，頸管和蓋塞完全，儲精提筒完好，盛裝液氮經過1d預冷並觀察其損耗率，各項指標合格後方可使用。

2. 填充液氮時要小心謹慎

　　對於新罐或處於乾燥狀態的罐一定要緩慢填充並進行預冷，以防降溫太快損壞內膽，減少使用年限。或者將液氮運輸罐的液氮經漏斗注入儲存罐內，為了防止液氮飛濺，可在漏斗內襯一塊紗布。

3. 及時補充液氮

　　當液氮消耗掉1/2時，應立即補充液氮。罐內液氮的剩餘量可用稱量法來估算，也可用帶刻度的木尺或細木條等插至罐底，經10s後取出，通過測量結霜的長度來估算。

4. 液氮罐的放置

　　液氮罐需放置在陰涼、通風且乾燥的室內，不得曝曬，不可橫倒放置。

5. 做好定期的清洗和保養工作

　　每年應清洗一次罐內雜物，將空罐放置2d後，用40～50℃中性洗滌劑擦洗，再用清水多遍沖洗，乾燥後方可使用。使用過程中，如發現罐的外壁結霜，說明罐的夾層功能失去作用，要盡快轉移精液。

medium-sized liquid nitrogen tanks have a coat and a belt for carrying the back.

　　In general, liquid nitrogen tank should have good insulation performance, durable and easy to use, so as to better meet the needs of production.

7.3　Use of liquid nitrogen tank

7.3.1　Check carefully before use

　　The newly purchased or long-term unused liquid nitrogen tank must be free from damage and abnormality on the outside, dry inside without foreign matters, complete neck tube and cap plug, and intact lifting cylinder. The liquid nitrogen can be used only after it is precooled for one day and all indicators are qualified.

7.3.2　Be careful when filling liquid nitrogen

　　For the new tank or the tank in dry state, it must be filled slowly and precooled to prevent the tank from being damaged and reduce the service life. In order to prevent liquid nitrogen from splashing, a piece of gauze can be lined in the funnel.

7.3.3　Replenish liquid nitrogen in time

　　When half of the liquid nitrogen is consumed, it should be replenished immediately. The residual liquid nitrogen in the tank can be estimated by weighing, or inserted into the bottom of the tank with a calibrated wooden ruler or sliver, then taken out after ten seconds, and estimated by measuring the frosting length.

7.3.4　Placement of liquid nitrogen tank

　　The liquid nitrogen tank should be placed in a cool, ventilated and dry room, and should not be exposed to the sun or placed horizontally or upside down.

7.3.5　Regular cleaning and maintenance

　　The sundries in the tank should be cleaned once a year. After the empty tank is placed for two days, it should be scrubbed with 40-50℃ neutral detergent, then washed with clean water for several times. It can be used after drying. In the process of use, if frost is found on the outer wall of the tank, it means that the vacuum of the jar is out of order and semen should be transferred as soon as possible.

6. 防凍傷

液氮是一種超低溫液體，如濺到皮膚上會引起凍傷，因此在灌充和取出液氮時應注意做好自身的防護。

7.3.6 Frostbite

Liquid nitrogen is a kind of ultra-low temperature liquid. If it is splashed on the skin, it will cause frostbite similar to burn. Therefore, we should pay attention to self-protection when filling and taking out liquid nitrogen.

任務 1　精液的品質評定
Task 1　Evaluation of Semen

任務描述

有一公豬，一次採集精液 285mL，精液呈乳白色、略有腥味，精子活力為 0.78，精子密度為 2.4 億個/mL，精子畸形率為 12.39%。試分析該公豬的精液品質是否合格？

Task Description

In a boar, 285 mL semen was collected at one time. The semen was milky white and smelt slightly fishy. Sperm viability rate was 0.78, density was 240 million/mL, and sperm abnormality rate was 12.39%. Please analyze whether the semen of this boar is qualified?

任務實施

一、精液的感官評定

（一）準備工作

將採集好的精液做好標記，迅速置於 37℃ 左右的溫水或保溫瓶中備用。

（二）檢查方法

1. 採精量

採精後應立即檢測採精量。將採集的精液盛放在帶有刻度的集精杯或試管中，直接讀出精液量。

注意：豬、馬的精液需濾除膠狀物質後再檢測。

2. 顏色

肉眼觀察裝在透明容器中的精液顏色。

Task Implementation

1　Sensory evaluation of semen

1.1　Preparations

The collected semen should be labeled and immediately placed in warm water at around 37℃ or thermos flask.

1.2　Methods

1.2.1　Volume

The volume of semen should be detected immediately after semen collection. The semen should be placed in a cup or a test tube with scales to directly read the volume of semen.

Note：The semen of boar and male horse should be filtered before detecting the volume of semen.

1.2.2　Color

Observe the color of the semen contained in a transparent container.

3. 氣味

用手慢慢在裝有精液的容器上方搧動，並嗅聞精液的氣味。

4. 雲霧狀

觀察裝在透明容器中的精液狀態，主要觀察液面的變化情況。

（三）結果評定

1. 採精量

採精量的多寡受多種因素影響，但超出正常範圍（表 2-2）太多或太少，應及時查明原因。

1.2.3 Odor

Slowly wave plam of the hand over the container and smell the odor of the semen.

1.2.4 Cloudy appearance

Observe the semen state in a transparent container, mainly observe the changes of the fluid.

1.3 Result evaluation

1.3.1 Volume

The volume of semen is influenced by many factors, but if it exceeds the normal range (Table 2-2) by too much or too little, the reasons should be found out in time.

表 2-2　成年公畜（禽）的採精量
Table 2-2　Semen volume of livestock（poultry）

動物種類 Animal species	一般採精量/mL General volume/mL	正常採精範圍/mL Normal range/mL
乳牛 Dairy cattle	5～10	0.5～14
肉牛 Beef cattle	4～8	0.5～14
水牛 Buffalo	3～6	0.5～12
山羊 Goat	0.5～1.5	0.3～2.5
綿羊 Sheep	0.8～1.2	0.5～2.5
豬 Pig	150～300	100～500
馬 Horse	40～70	30～300
雞 Rooster	0.5～1.0	0.2～1.5

2. 顏色

正常精液一般為乳白色或灰白色。若精液顏色異常，應當廢棄，並查明原因，及時治療。

3. 氣味

正常精液略帶腥味，牛、羊精液除具有腥味外，另有微汗脂味。如有異常氣味，可能是混有尿液、膿汁、糞渣或其他異物，應廢棄。

4. 雲霧狀

精液的液面呈上下翻滾狀態，像雲霧一樣，稱為雲霧狀。雲霧狀越明顯，說明精液密度越大，活力越高。正常未

1.3.2 Color

The color of normal semen is usually milky white or greyish white. If the semen color is abnormal, it should be discarded. We should find out the cause and treat in time.

1.3.3 Odor

Normal semen has a slightly fishy odor, the semen of bull and ram have a slightly sweaty and fatty odor. If there is abnormal odor, it could be mixed with urine, pus, feces or others, and should be discarded.

1.3.4 Cloudy appearance

The fluid of semen rolls up and down like a cloud, so it is called cloudy appearance. The more obvious cloudy appearance of semen is, the higher

稀釋的牛、羊精液肉眼可觀察到雲霧狀。

二、精子活率評定

(一) 準備工作

(1) 器械準備。光電顯微鏡，清洗乾淨的載玻片和蓋玻片，移液槍，吸頭，擦鏡紙。有條件的亦可選用全自動精子分析儀。

(2) 試劑準備。生理鹽水。

(3) 精液準備。新鮮精液。

(二) 評定方法

用移液槍吸取1滴原精液或經生理鹽水稀釋的精液，滴在載玻片上，呈45°蓋好蓋玻片，載玻片與蓋玻片之間應充滿精液，避免氣泡存在，置於顯微鏡下觀察，估測呈直線運動的精子數占總精子數的百分率。

(三) 結果評定

精子活率是指精液中呈直線運動的精子數占總精子數的百分率。評定精子活率多採用「十級評分制」法，劃分為1.0、0.9、0.8、…0.2、0.1等10個等級。如果視野中100％的精子作直線前進運動，活率評為1.0級；90％者評為0.9，80％者評為0.8，依此類推。各種動物的新鮮精液的精子活率一般為0.7～0.8，否則生產上不能用於保存和輸精。

(四) 注意事項

(1) 牛、羊和雞的精子密度較大，可用生理鹽水稀釋後再檢查。

(2) 精子活率是評價精液品質的一個重要指標，與受精力密切相關，一般在採精後、精液處理前後及輸精前都要進行檢測。

density of semen is and the higher viability of sperm is. Cloudy appearance in undiluted semen of bull and ram can be seen directly.

2 Evaluation of sperm viability

2.1 Preparations

(1) Equipment. Microscope, clean glass slides and coverslips, pipetting gun, sucker, lens wipes. A fully automatic sperm analyzer can also be used if conditions permit.

(2) Reagents. Normal saline.

(3) Semen. Fresh semen.

2.2 Methods

A drop of the original semen or diluted semen is dropped on the glass slide with a pipetting gun, covered by coverslip at an angle of 45°. The space between slide and coverslip should be filled with semen to avoid bubbles. Observe under a microscope and estimate the percentage of sperm in linear motion.

2.3 Result evaluation

Sperm viability refers to the percentage of sperm in linear motion. Sperm viability is assessed by ten-grade scoring system, which is divided into 10 grades such as 1.0, 0.9, 0.8, … 0.2, 0.1. If 100% sperm moves in a straight line, sperm viability is 1.0, 90% sperm moves in a straight line is 0.9, and so on. Viability of fresh semen in various animals is generally 0.7-0.8, otherwise, it can not be used for preservation and insemination in production.

2.4 Cautions

(1) The sperm density of bull, ram and cook is high, which can be diluted with normal saline and then examined.

(2) Sperm viability is an important index for evaluating semen quality, which is closely related to fertilization ability. Generally, it should be tested after semen collection, before and after semen treatment and before insemination.

（3）溫度對精子活率影響較大，要求檢查溫度為 37～38℃，如果沒有保溫裝置的，檢查速度要快，在 10s 內完成。

（4）精子活率評定帶有一定的主觀性，應觀察 3～5 個視野，取平均值。如果採用電視顯微鏡，可幾人同時觀察，評定結果較為準確。

三、 精子密度評定

精子密度又稱精子濃度，是指每毫升精液中所含有的精子數。目前，常用的評定方法有估測法、血球計數法和精子密度儀測定法。

（一）估測法

1. 準備工作

（1）器材準備。顯微鏡、載玻片、蓋玻片、移液槍、吸頭、擦鏡紙等。

（2）精液的準備。新鮮的精液。

2. 評定方法

通常結合精子活率評定進行。取 1 小滴精液滴於清潔的載玻片上，蓋上蓋玻片，使精液分散成均勻的薄層，置於顯微鏡下觀察精子間的空隙。

3. 結果評定

根據顯微鏡下精子的密集程度，把精子密度大致分為密、中、稀三個等級（圖 2-5）。

（3）Temperature has a great influence on sperm viability. The evaluation temperature should be about 37-38℃. If there is no heat preservation device, the evaluation should be do quickly and completed in 10 seconds.

（4）The evaluation of sperm viability is subjective. We should observe 3-5 fields of view and take the average. If a TV microscope is used, several people can observe at the same time, and the evaluation results are more accurate.

3 Evaluation of sperm density

Sperm density, also known as sperm concentration, refers to the number of sperms contained per milliliter of semen. At present, the commonly used evaluation methods are estimation method, blood count method and sperm density measurement method.

3.1 Estimation method

3.1.1 Preparations

（1）Assess equipment. Microscope, glass slides, coverslips, pipetting gun, sucker, lens wipes, etc.

（2）Semen. Fresh semen.

3.1.2 Methods

This method is usually combined with sperm viability assessment. Take a drop of semen onto a cleaned glass slide, then covered with coverslip. In this process, make sure the semen disperses into a uniform thin layer and observe the gap between sperms under the microscope.

3.1.3 Result evaluation

According to the degree of sperm concentration under the microscope, the sperm density can be roughly divided into three grades: dense, medium and sparse(Figure 2-5).

由於各種動物精子密度差異很大，很難使用統一的等級標準，檢查時應根據經驗，對不同的動物採用不同的標準（表 2-3）。該法具有較大的主觀性，誤差也較大，但簡便易行，在基層人工授精站常採用。

The sperm density of various animals is quite different, so it is difficult to use a unified grade standard. Different standards should be used for different animals according to experience (Table 2-3). The method has great subjectivity and large error range, but it is simple and easy, and is often used in local artificial insemination stations.

密 Dense　　中 Medium　　稀 Sparse

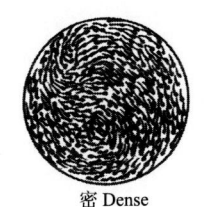

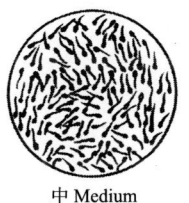

圖 2-5　精子密度

Figure 2-5　Sperm density

表 2-3　各種動物精子密度

Table 2-3　Sperm density of various animals

動物類別 Animal species	精子數/（億個/mL） Sperm concentration/（100 million/mL）		
	密 Dense	中 Medium	稀 Sparse
牛 Cattle	>15	10~15	<10
羊 Sheep	>25	20~25	<20
豬 Pig	>3	1~3	<1
雞 Rooster	>40	20~40	<20

(二) 血球計數法

1. 準備工作

（1）器材準備。顯微鏡、血球計數板、蓋玻片、移液槍、吸頭、計數器、擦鏡紙等。

（2）試劑準備。3% NaCl 溶液。

（3）精液準備。新鮮的精液。

2. 評定方法

（1）清洗器械。將血球計數板及蓋玻片用蒸餾水沖洗，使其自然乾燥。

（2）稀釋精液。用 3% NaCl 溶液對精液進行稀釋，稀釋倍數以方便計數

3.2　Blood count method

3.2.1　Preparations

(1) Assess equipment. Microscope, hemacytometer, coverslips, pipetting gun, sucker, lens wipes, etc.

(2) Reagents. 3% NaCl solution.

(3) Semen. Fresh semen.

3.2.2　Methods

(1) Clean the hemocytometer and coverslips with distilled water, then allow them to dry naturally.

(2) The semen is diluted with 3% NaCl solution, and the dilution factor should be made convenient for count-

為準。牛、羊的精液一般稀釋100、200倍，豬的精液稀釋10、20倍。

（3）找準方格。將血球計數板置於載物臺上，蓋上蓋玻片，先在100倍顯微鏡下查看方格全貌（圖2-6），再在400倍顯微鏡下查找其中一個中方格。

（4）鏡檢。將稀釋的精液滴一滴於計數室上蓋玻片的邊緣，使精液自動滲入計算室（圖2-7），靜置3min，計數具有代表性的5個中方格內的精子數。一般計數四個角的中方格和中間一個中方格（或對角線上的五個中方格）。

（5）計算。1mL原精液的精子數＝5個中方格內的精子總數×5×10×1 000×稀釋倍數。

ing. The semen of cattle and sheep is diluted 100 or 200 factors, and the semen of pig is diluted 10 or 20 factors.

(3) Place the hemacytometer on the objective table, then cover the coverslip. First, observe the full view of the counting chamber under low magnification ($100\times$)(Figure 2-6), then look for one of the large double-ruled squares under a high magnification (approximately $400\times$).

(4) Diluted semen is dropped on the edge of the coverslip, then the semen is infiltrated automatically into the counting chamber (Figure 2-7). Wait about 3 minutes before starting to count the number of sperm in five representative squares. Generally, the middle medium square and four medium squares in the corner (or five medium squares of the diagonal) are counted.

(5) The number of sperm per milliliter is equivalent to the number of sperm counted over 5 double-ruled squares $\times 10\ 000 \times$ the dilution factor.

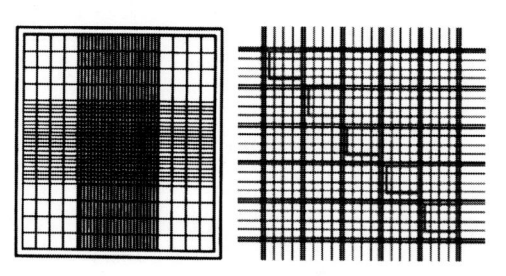

圖 2-6　計數室結構

Figure 2-6　Structure of counting chamber

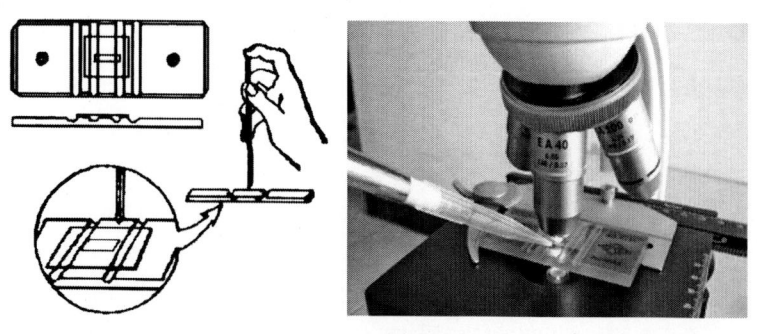

圖 2-7　滴加精液

Figure 2-7　Dripping semen

3. 結果評定

正常情況下，雞的精子密度較大，為 20 億～40 億個/mL；羊為 20 億～30 億個/mL；牛為 10 億～15 億個/mL；豬為 1 億～3 億個/mL。該方法因檢測速度慢，在生產上用得較少，但結果準確，一般用於結果的校準及產品品質的檢測。

4. 注意事項

（1）滴入精液時，不要使精液溢出蓋玻片，也不可因精液不足而導致計數室內有氣泡或乾燥之處。

（2）計數時，以頭部壓線為準，按照「數頭不數尾、數上不數下、數左不數右」的原則，避免重複或漏掉（圖 2-8）。

（3）為了減少誤差，應連續檢查兩次，取其平均值，若兩次計數誤差大於 10％，則應做第三次檢查。

3.2.3 Result evaluation

The sperm density of rooster is relatively large, which is 2 billion to 4 billion per milliliter. It is 2 billion to 3 billionin in ram, 1 billion to 1.5 billion in bull, 100 million to 300 million in boar. Because of its slow detection speed, this method is seldom used in the process of production, but the result is accurate, and is generally used for the calibration of result and the detection of product quality.

3.2.4 Cautions

(1) When dropping semen, do not let the semen overflow the coverslip, nor cause bubbles or dry areas in counting chamber due to insufficient semen.

(2) To avoid counting twice any sperm that are at the boundary of adjacent double-ruled squares, routinely count only those boundary sperm head on or within the top and left edges of each double-ruled squares (Figure 2-8).

(3) In order to reduce errors, two consecutive checks should be made and take their average values. If the two counting errors are greater than 10%, a third check should be made.

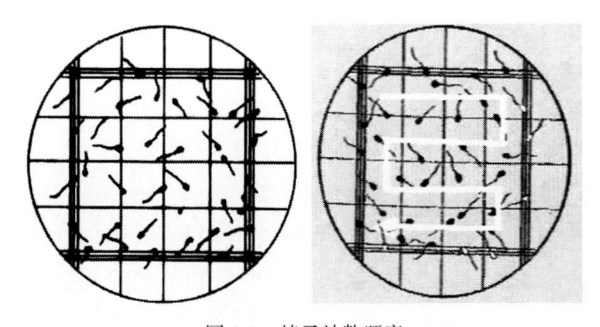

圖 2-8　精子計數順序

Figure 2-8　The order of sperm counting

（三）精子密度儀測定（圖 2-9）

將待檢精液樣品按一定比例稀釋，置於精子密度儀中讀取結果。此法快速、準確、操作簡便，廣泛用於畜禽的精子密度測定。

3.3 Sperm density measure (Figure 2-9)

The semen samples are diluted in a certain proportion and placed in the sperm densitometer to read the results. This method is rapid, accurate and easy to operate. It is widely used for the determination of sperm density in livestock and poultry.

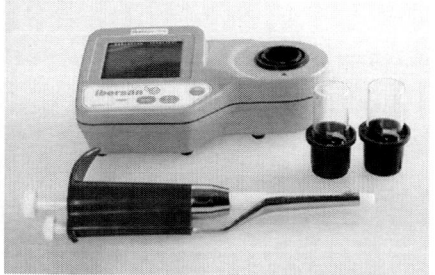

圖 2-9　精子密度儀測定
Figure 2-9　Sperm density measure

四、精子畸形率評定

(一) 準備工作

(1) 器械準備。顯微鏡、載玻片、移液槍、吸頭、計數器、染色缸、擦鏡紙。

(2) 試劑準備。藍墨水（或紅墨水或 0.5％龍膽紫溶液）、96％酒精等。

(3) 精液準備。新鮮的精液或冷凍保存的精液。

(二) 評定方法

(1) 抹片。用移液槍吸取精液 1 滴，滴於載玻片一端，以另一載玻片的頂端呈 35°角抵於精液滴上，精液呈條狀分布在兩個載玻片接觸邊緣之間，自右向左移動，將精液均勻塗抹於載玻片上（圖 2-10）。

(2) 乾燥。抹片於空氣中自然乾燥。

(3) 固定。置於 96％酒精固定液中固定 5～6min，取出沖洗後陰乾。

(4) 染色。用藍（紅）墨水染色 3～5min，用緩慢水流沖洗乾淨並使之乾燥。

(5) 鏡檢。將製好的抹片置於 400 倍顯微鏡下，查數不同視野的 300～

4　Evaluation of sperm abnormality

4.1　Preparations

(1) Assess equipment. Microscope, clean glass slides, pipetting gun, sucker, hand tally counter, dyeing tanks, lens wipes.

(2) Reagents. Blue ink (or red ink or 0.5% gentian violet solution), 96% alcohol, etc.

(3) Semen. Fresh semen or frozen semen.

4.2　Methods

(1) Smear. A drop of semen is dropped on one end of a glass slide with a pipetting gun. The top of the other glass slide is contacted to the semen drop at a 35° angle. The semen drop is distributed in a strip between the contact edges of the two glass slides. Then move the top glass slide from right to left and spread the semen evenly on the glass slide below it (Figure 2-10).

(2) Drying. The smear dries naturally in the air.

(3) Fixation. Fixed in 96% alcohol for 5-6 minutes. Take it out and rinse, then dry it in the shade.

(4) Staining. Dye with blue (red) ink for 3-5 minutes, rinse with slow streaming water then dry it.

(5) Microscopic examination. The smear is observed under microscope (400×), and 300-500 sperms

500個精子，記錄畸形精子的數量，並計算精子畸形率。

are counted from different views. Then record the number of abnormal sperms and calculate the rate of abnormality.

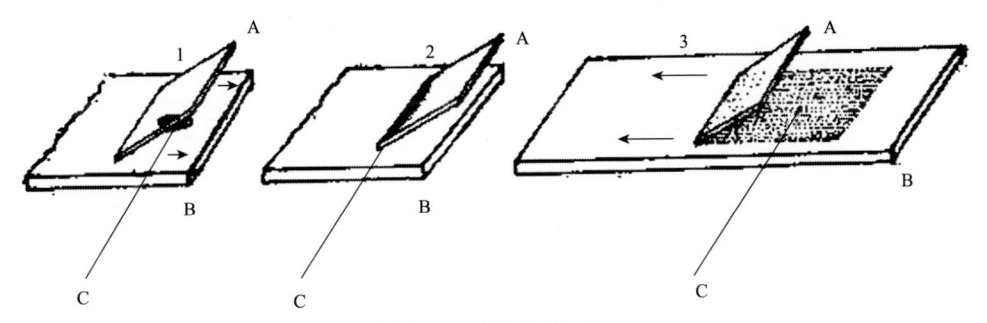

圖 2-10　精液抹片示意
Figure 2-10　Semen smear

1. A 片後退，使其邊緣接觸精液小滴（C）　2. 精液均勻分散在 A 片的邊緣
3. A 片向前推進，使精液均勻塗抹在 B 片上

1. Glass slide A is contacted to the semen drop（C）
2. The semen drop is distributed in a strip between the contact edges of glass slide A
3. Move the glass slide A from right to left and spread the semen evenly on the glass slide B

$$精子畸形率（Abnormality\ rate）= \frac{畸形精子數（No.\ of\ abnormal\ sperms）}{精子總數（No.\ of\ sperms\ recorded）} \times 100\%$$

（三）結果評定

凡形態和結構不正常的精子統稱為畸形精子。畸形精子類型很多，按其形態結構可分為三類（圖 2-11）：頭部畸形，如頭部巨大、瘦小、細長、缺損、雙頭等；頸部畸形，如頸部膨大、纖細、曲折、雙頸等；尾部畸形，如尾部膨大、纖細、彎曲、曲折、迴旋、雙尾等。

正常情況下，精液中會含有一定比例的畸形精子，一般牛、豬不超過 18%，羊不超過 14%，馬不超過 12%。

4.3　Result evaluation

Sperms with abnormal morphology and structure are called abnormal sperm. There are many types of abnormal sperm, which can be basically divided into three types according to their morphology and structure（Figure 2-11）: head abnormalities, such as large, small, slender, broken, and double-head; neck abnormalities, such as intumescent, slender, tortuous, and double-neck; tail abnormalities, such as intumescent, slender, tortuous, twisted, circling, double-tail.

Usually, there is a certain proportion of abnormal sperm in semen, it is no more than 18% in bull and boar, no more than 14% in ram and no more than 12% in male horse.

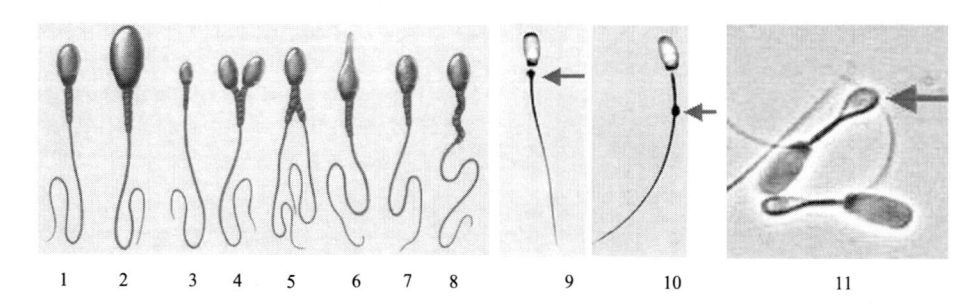

圖 2-11　畸形精子類型

Figure 2-11　Types of abnormal sperm

1. 正常精子　2. 頭部膨大　3. 頭部瘦小　4. 雙頭精子　5. 雙頸　6. 頭部細長　7. 頭部破損
8. 頸部曲折　9. 近端原生質滴　10. 遠端原生質滴　11. 別針尾

1. normal sperm　2. large head　3. small head　4. double-head　5. double-neck　6. slender head　7. broken head
8. tortuous neck　9. proximal protoplasmic drop　10. distal protoplasmic drop　11. pin-shaped tail

任務 2　精液的稀釋
Task 2　Dilution of Semen

任務描述 / Task Description

精液稀釋是人工授精中的一個重要技術環節，只有經過稀釋的精液，才適於保存、運輸及輸精。精液的保存方法不同，稀釋的倍數及稀釋液的成分也不相同。如何正確稀釋精液？

Semen dilution is an important technical link in artificial insemination. Only diluted semen is suitable for preservation, transportation and insemination. The methods of semen preservation are different, the dilution factors and the composition of the dilution are also different. How to dilute semen correctly?

任務實施 / Task Implementation

一、確定稀釋倍數

精液稀釋前需要確定稀釋倍數，稀釋倍數主要根據精子密度、稀釋液的種類以及保存時間的長短來定。一般牛精液的稀釋倍數為 10～40 倍，豬精液為 2～4 倍。

1　Determine the dilution factor

Before semen dilution, it is necessary to determine the dilution factor, which is mainly determined by the sperm density, the type of dilution solution and the length of storage time. The dilution factor of bovine semen is 10-40 times, and that of boar semen is 2-4 times.

二、配製稀釋液

1. 準備工作

將玻璃器皿清洗乾淨並晾乾,然後用錫紙將瓶口封好置於 120℃恆溫乾燥箱乾燥 1h,冷卻備用。

2. 配製稀釋液

按照配方準確稱量藥品,放置於燒杯中,加入蒸餾水攪拌使其充分溶解,過濾溶液置於 100℃水浴鍋消毒 10～20min,冷卻後加入卵黃、抗生素、激素、維他命等成分,混合均勻。

三、正確稀釋精液

採精後應立即對精液進行稀釋,將稀釋液和精液置於 30℃左右環境進行同溫處理。稀釋時,將稀釋液沿著容器壁緩緩加入精液中,邊加入邊搖晃,要避免劇烈震盪。稀釋完畢後,應該立即進行精子活率檢查。如果精子活率沒有變化,則可進行分裝、保存。如果精子活率下降,說明稀釋液配製不當或稀釋處理不當,則應棄掉精液並查明原因。

四、注意事項

（1）配製稀釋液的器具必須徹底清洗,嚴格消毒。

（2）選用新鮮的蒸餾水。

（3）藥品試劑要純淨。

（4）乳和乳粉需新鮮。將一定量的鮮乳或充分溶解的乳粉溶液,用脫脂棉進行過濾,再用水浴鍋加熱到 92～95℃消毒滅菌 10min,取出降溫,除去奶皮備用。

2 Preparation of diluent

2.1 Preparations

The glassware should be cleaned and dried, then the bottle mouth should be sealed with tin foil and placed in a constant temperature drying oven at 120℃ for one hour, and then cooled for use.

2.2 Prepare the diluent

Drugs should be accurately weighed in beaker according to the formula, stirred with distilled water to dissolve fully, then the filtered solution should be disinfected in a water bath at 100℃ for 10-20 minutes. After cooling, egg yolk, antibiotics, hormones, vitamins and other ingredients are added and mixed well.

3 Correct dilution of semen

After semen collection, the semen should be diluted immediately, and the diluent and semen are treated at the same temperature at about 30℃. During dilution, the diluent is slowly added to semen along the wall of the container and gently shake it while adding, severe shocks should be avoid. Sperm viability should be examined immediately after dilution. If the sperm viability does not change, it can be packaged and preserved. If the sperm viability decreases, which indicates that the dilution is improperly prepared or diluted, the semen should be discarded and the cause should be identified.

4 Cautions

（1）Appliances for preparing diluents must be thoroughly cleaned and strictly disinfected.

（2）To choose fresh distilled water.

（3）Drugs should be purified.

（4）Milk and milk powder must be fresh. A certain amount of fresh milk or fully dissolved milk powder solution is filtered with absorbent cotton, and then heated to 92-95℃ for 10 minutes in a water bath. Before using it, the temperature should be cooled down and the milk skin should be removed.

（5）雞蛋要新鮮，卵黃中不應混入蛋白和卵黃膜。經加熱消毒過的稀釋液，待溫度降至40℃以下再加入卵黃，並注意充分溶解。

（6）抗生素、酶類、維他命、激素等需在稀釋液加熱滅菌後，溫度降至40℃以下再加入。

（7）稀釋液最好現用現配。

(5) Eggs should be fresh and egg yolk should not be mixed with protein and yolk membrane. Before adding the egg yolk, the temperature should be dropped below 40℃ after heating and disinfection. The egg yolk must be dissolved sufficiently after adding.

(6) After the diluent is heated and sterilized, and the temperature drops below 40℃, antibiotics, enzymes, vitamins, hormones can be added.

(7) After the diluent is prepared, it is better to be used soon.

任務 3　冷凍精液的製作
Task 3　Production of Frozen Semen

任務描述 / Task Description

冷凍精液是利用液氮作為冷源，將精液經過特殊處理後浸泡在液氮中，可以進行長期保存和遠距離運輸。冷凍精液的劑型主要有顆粒凍精和細管凍精兩種，後者不易污染、便於標記、適於機械化生產，解凍和輸精也比較方便，生產上普遍使用。牛的細管凍精多用 0.25mL 劑型。大家熟悉牛冷凍精液的製作流程嗎？

Frozen semen is to immerse specially treated semen in liquid nitrogen for long-term preservation and long-distance transportation. The types of frozen semen mainly include granular type and straw type. The latter is not easy to contaminate, easy to label, convenient for mechanized production, thawing and insemination, and is widely used in production. The bovine frozen semen mostly uses the size of 0.25 mL straws. Are you familiar with the production process of frozen semen?

任務實施 / Task Implementation

一、採精及精液品質檢查

用假陰道採集公牛的精液，立即進行品質檢查，一般要求精子活率不低於 0.7，精子密度不少於 8 億個/mL，精子畸形率不超過 15%。

1　Semen collection, assessment and dilution

After the semen of the bull is collected with the artificial vagina, quality assessment should be carried out immediately. Generally, the sperm viability is not less than 0.7, the sperm density is not less than 800 million/mL, and the sperm abnormality is not more than 15%.

專案二 精液的處理
Project Ⅱ Semen Treatment

二、 精液稀釋

經品質檢查後，及時對精液進行稀釋。精液稀釋採取緩慢、少量添加的方法，切忌快速、多量稀釋和劇烈震盪。

1. 一次稀釋法

將配製好的含有卵黃、甘油的稀釋液按一定比例加入精液中，適用於低倍稀釋。

2. 兩次稀釋法

為了縮短甘油與精子的接觸時間，常採用兩次稀釋法。首先用不含甘油的稀釋液Ⅰ和精液置於30℃環境進行同溫處理，按稀釋倍數進行半倍稀釋；然後把稀釋精液連同稀釋液Ⅱ一起緩慢降溫至0～5℃，在此溫度下進行第二次稀釋。

三、 精液分裝

已經稀釋的精液，通過電腦控制在一體機上進行印刷、灌裝及封口。細管凍精的劑型有0.25mL、0.5mL和1.0mL，國際多採用0.25mL劑型（圖2-12）。管壁外影印上產地、公牛品種、公牛號、製造日期等資訊。

2 Semen dilution

After quality assessment, the semen should be diluted in time. Sperm dilution should be done slowly and in a small amount. Don't do it hastily or dilute in a large amount and shock violently.

2.1 One-step dilution method

The prepared dilution containing egg yolk and glycerin is added to the semen in a certain ratio, which is suitable for low-dilution.

2.2 Two-step dilution method

In order to shorten the contact time of glycerin with sperm, two-step dilution method is often used. First, the glycerol-free diluent Ⅰ and semen are placed in a 30℃ environment for isothermal treatment, half-diluted according to the dilution factor. Then the diluted semen is slowly cooled to 0-5℃ together with the diluent Ⅱ, and the second dilution is made at this temperature.

3 Semen packaging

The diluted semen is printed, filled and sealed on the integrated machine controlled by a computer. There are several sizes of straws such as 0.25 mL, 0.5 mL, and 1.0 mL, and the 0.25 mL straw is used more frequently in the world (Figure 2-12). Outside of straw wall information such as origin, breed, bull number, production date are printed.

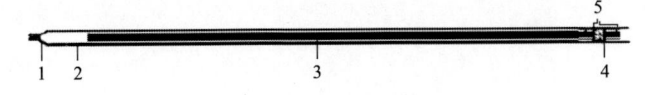

圖 2-12　細管凍精結構示意
Figure 2-12　The structure of frozen semen straws
1. 超音波封口　2. 空氣柱　3. 精液柱　4. 聚乙烯醇粉　5. 棉塞
1. ultrasonic sealing　2. air column　3. semen column　4. polyvinyl alcohol powder　5. tampon

四、 降溫平衡

將分裝好的細管精液置於0～5℃環境中平衡2.5～4h，使甘油充分滲入

4 Cooling and balancing

The tubular semen is balanced for 2.5-4 hours at 0-5℃, so that glycerol could penetrate into the sperm

精子內部，起到抗凍保護作用。

五、凍結

將平衡後的細管精液平鋪在冷凍板上，距液氮面 1～2cm 處燻蒸 5～10min，然後將合格的細管凍精移入液氮罐內保存。也可利用冷凍儀製作細管凍精。

六、保存

目前，生產中普遍採用液氮為冷源，液氮罐作為儲存凍精的容器。根據液氮的揮發情況，要定期往液氮罐中添加液氮。

七、解凍及品質評定

燒杯中盛滿 38℃ 左右溫水，打開液氮罐，把鑷子放至罐口預冷，提起提筒，迅速夾取一支凍精，放入燒杯中，並輕輕攪拌 10s 左右（圖 2-13）。待凍精融化後剪去細管的封口端，裝入細管輸精槍中。細管凍精品質檢查可按批抽樣評定，精子活率應不低於 0.35。

and plays a role of anti-freezing.

5 Freezing

The balanced tubular semen is laid flat on the freezing plate and fumigated for 5-10 minutes at 1-2 cm above the liquid nitrogen surface. Then the qualified tubular frozen semen could be transferred to the liquid nitrogen container for preservation. The freezer can also be used to freeze semen.

6 Storage

At present, liquid nitrogen is widely used as cold source in production, and liquid nitrogen container is used for storing frozen semen. According to the volatilization, liquid nitrogen should be added to the container periodically.

7 Thawing and quality assessment

Fill the beaker with warm water at about 38℃, open the liquid nitrogen container, put the tweezers upon the container to pre-cool, lift the lifting barrel, quickly grab a frozen tubule, put it into the beaker, and gently stir for about 10s (Figure 2-13). After the frozen semen is dissolved, the sealing end of the tubule should be cut and loaded into an insemination gun. The quality of frozen semen can be assessed by batch sampling, and the viability should not be less than 0.35.

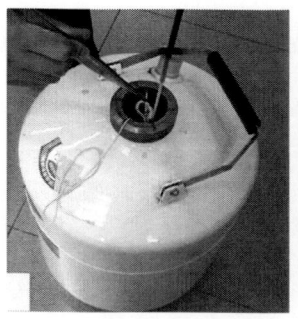

圖 2-13　細管凍精的解凍

Figure 2-13　Thawing of frozen semen from tubules

任務 4　精液的保存與運輸
Task 4　Preservation and Transportation of Semen

任務描述 / Task Description

精液保存的目的是為了延長精子在體外的存活時間，便於長途運輸，從而擴大精液的使用範圍。精液保存的方法，按保存溫度不同可分為常溫保存、低溫保存和冷凍保存三種。對於遠距離購買精液的豬場，運輸過程至關重要。將精液從甲地運往乙地，應如何運輸？

The purpose of semen preservation is to prolong the survival time of sperm in vitro, facilitate long-distance transport, and expand the scope of semen use. Semen preservation methods can be divided into room temperature preservation, low-temperature preservation and cryopreservation. Transportation is critical for pig farms that purchase semen over long distances. How should semen be transported from place A to place B?

任務實施 / Task Implementation

一、豬精液的保存與運輸
1　Preservation and transportation of boar semen

（一）精液的分裝
1.1　Semen packaging

精液稀釋後按每頭母豬一次的輸精量進行灌裝。精液分裝常採用瓶裝和袋裝兩種。瓶裝的精液分裝簡單方便，易於操作，容量最高為 100mL（圖 2-14）；袋裝的精液分裝一般需要專門的精液分裝機，用機械分裝、封口，容量一般為 80mL（圖 2-15）。

After the semen is diluted, it is packaged according to the amount of insemination in each sow. Semen packaging is often used both bottles and bags. Bottled semen is easy to install and handle. The bottled semen volume is up to 100 mL (Figure 2-14). The bagged semen is packed generally with a special semen packer, which is mechanically packed and sealed. The semen volume is typically 80 mL (Figure 2-15).

分裝後的精液，要逐個黏貼標籤。為了便於區分，一般一個品種一個顏色。分好後將精液瓶加蓋密封，封口時盡量擠出瓶中空氣，貼上標籤，標明公豬的品種、耳號、採精日期、保存有效期、生產單位及相應的使用說明。

After packaging semen, the label should be affixed one by one, in order to distinguish easily, generally one breed has one color. When sealing, the air in the bottle should be squeezed out, and the label should have the information of boar's breed, ear number, date of collection, expiration date, manufacturer and corresponding instructions.

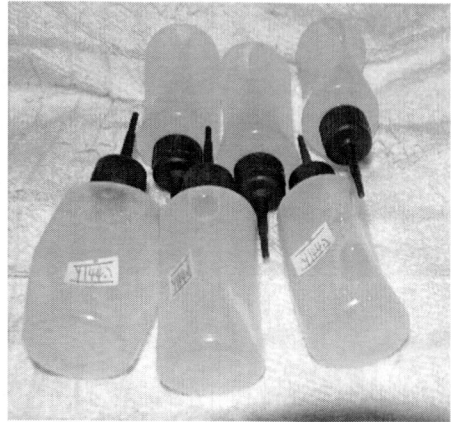

圖 2-14　分裝後的瓶裝精液
Figure 2-14　Bottled semen

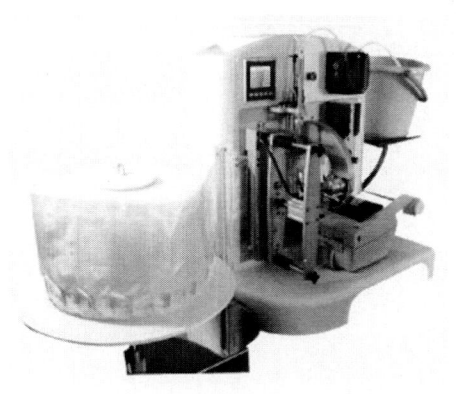

圖 2-15　豬精液自動灌裝影印一體機
Figure 2-15　Semen automatic filling and printing machine

(二) 精液的保存

分裝好的精液先置於 22～25℃ 的環境下放置 1～2h，然後移入 16～18℃ 的恆溫箱中保存（圖 2-16）。存放時，不論是瓶裝的還是袋裝的，均應平放，目的是為了增大精子沉澱後鋪開的面積，減少沉澱的厚度，降低精子死亡率。

1.2　Semen preservation

The packaged semen should be placed at 22-25℃ for 1-2 hours, and then stored in a thermostat at 16-18℃ (Figure 2-16). When storing, whether it is bottled or bagged, it should be laid flat in order to increase the area of sperm precipitation, and reduce the thickness of sperm and the rate of sperm death.

專案二　精液的處理
Project Ⅱ　Semen Treatment

圖 2-16　豬精液保存恆溫箱
Figure 2-16　Thermostat for boar semen preservation

在保存期內要注意三點：一是盡量減少恆溫箱門的開關次數，防止頻繁改變溫度對精子的打擊；二是每隔12h輕輕翻動一次，以防止精子沉澱；三是每天檢查恆溫箱內溫度計的變化，防止溫度出現明顯的波動。若保存過程中出現停電現象，應全面檢查儲存的精液品質。精液保存時間的長短，因稀釋液成分的不同而異，一般可保存2～3d。

（三）精液的運輸

精液運輸是豬場人工授精順利進行的必要環節，具有提高種公豬利用率、防止疾病傳播、更新豬群血液等優點。豬精液運輸需注意以下幾點：

（1）運輸的精液需按規定進行稀釋和保存，有詳細的說明書，標明站名、公豬品種和編號、採精日期、精液劑量、稀釋倍數、精子活率和密度等。

（2）精液包裝要嚴密，不能發生泄

Pay attention to three points during the storage period. First, reduce the number of opening and closing of the thermostat door to prevent frequent changes in temperature from hitting the sperm. Second, gently flip once every 12h to prevent sperm precipitation. Third, check the changes of thermometer in thermostat every day to prevent significant fluctuations in temperature. If there is a power outage during storage, semen quality should be checked comprehensively. The duration of semen preservation varies with the composition of the diluent. Generally, semen can be stored for 2-3 days.

1.3　Semen transportation

Semen transportation is an essential link for the successful artificial insemination in pig farms. It has the advantages of improving the utilization rate of boars, preventing the spread of diseases and refreshing the group blood. Pay attention to the following points when transporting boar semen.

(1) The transported semen should be diluted and preserved according to regulations. Detailed specifications include the station name, boar breed and number, date of semen collection, semen volume, dilution factor, sperm viability and density.

(2) Semen packaging should be tight to avoid

漏，運輸多採用精液運輸箱（圖 2-17）或廣口保溫瓶。

（3）盡量避免在運輸過程中產生劇烈震盪和碰撞。

（4）精液運輸過程中要注意保持溫度的恆定。

（5）精液運輸過程中要防止陽光直射到泡沫箱，更不能直射到輸精瓶上。

leakage, and the semen transport box (Figure 2-17) or wide-mouth thermos bottle is used for transportation.

(3) Try to avoid violent shocks and collisions during transportation.

(4) Pay attention to keeping the temperature constant during the transportation.

(5) During the transportation, it is necessary to prevent direct sunlight from reaching the foam box, and especially the bottle.

圖 2-17　豬精液運輸箱

Figure 2-17　Transport box of boar semen

二、牛精液的保存與運輸

（一）精液的保存

生產中多採用液氮罐儲存凍精。該法保存時間長，精液的使用不受時間、地域以及種公牛壽命的限制，對人工授精技術的推廣及現代畜牧業的發展均具有十分重要的意義。

取用凍精時，操作要敏捷迅速，將鑷子在液氮罐口預冷，凍精不可提出液氮罐口，凍精脫離液氮的時間不得超過 20s，取完後要迅速將餘下的凍精再次浸入液氮內，及時蓋上罐塞。在向另一個液氮儲存罐內轉移冷凍精液時，凍

2　Preservation and transport of bull semen

2.1　Semen preservation

Liquid nitrogen container is often used for storing frozen semenin. This method has a long preservation time, and the use of semen is not limited by time, region and life of the bull. It is of great significance for the promotion of artificial insemination and the development of modern animal husbandry.

When taking frozen semen, the operation should be quick and agile. Tweezers should be pre-cooled in the mouth of liquid nitrogen container. The frozen semen should not be put out of the liquid nitrogen container. The time for the frozen semen to leave the liquid nitrogen should not exceed 20 seconds. After taking the frozen semen, the rest of them should be im-

專案二 精液的處理
Project II Semen Treatment

在空氣中轉移時間不得超過 5s，並迅速將精液再次浸入液氮內。

（二）精液的運輸

由冷凍精液站提供的冷凍精液必須經過精子活率抽檢，查驗種公牛品種、公牛號及數量，與標籤一致無誤後方可運輸。盛裝精液的液氮罐必須確保其保溫性能，加外保護套或裝入液氮罐運輸箱（圖 2-18）內，罐與罐之間要用填充物隔開，在罐底加防震軟墊，防止顛簸撞擊，嚴防傾倒，裝卸車時要嚴防液氮罐碰擊，切不能在地上隨意拖拉，以免損壞液氮罐，降低使用壽命。運輸途中應隨時檢查並及時補充冷源。

mersed in the liquid nitrogen again quickly, and the mouth should be covered in time. When transferring frozen semen to another liquid nitrogen container, the frozen semen should not stay in the air for more than 5 seconds, and the semen should be quickly immersed in liquid nitrogen.

2.2 Semen transportation

The frozen semen provided by the frozen semen station must be sampled for sperm viability, and the bull breed shall be checked, along with the bull number and number of frozen semen, and it can be transported only if it is in accordance with the label. The liquid nitrogen container must ensure its thermal insulation performance, outer protective jacket should be added or be packed in the liquid nitrogen container transport box (Figure 2-18). The containers should be separated by fillers, and the bottom of the container should be shockproof. Prevent bumps and impacts, avoid tumbling, prevent the liquid nitrogen container from being hit, never drag on the ground when loading and unloading, so as not to damage the liquid nitrogen container and reduce the service life. The liquid nitrogen should be inspected and replenished at any time during transportation.

圖 2-18　液氮罐運輸箱
Figure 2-18　Transport box of liquid nitrogen container

專案三　發情鑑定與輸精
Project Ⅲ　Estrus Identification and Insemination

專案導學

　　發情鑑定是動物繁殖技術中最為基礎和關鍵的技術。通過掌握雌性動物的生殖器官構造和機能、卵泡的發育規律、雌性動物的發情生理及生殖激素對發情的調節等相關知識，做好發情鑑定，確定最佳的配種時間，規範完成輸精工作，提高受配率和受胎率。

學習目標

>>> 知識目標

• 理解雌性動物生殖器官的組成和主要生理功能。

• 掌握生殖激素的概念、分類及應用。

• 掌握雌性動物的發情生理和規律。

• 了解卵泡的發育和排卵機理。

• 理解生殖激素對發情週期的調節機理。

>>> 技能目標

• 能準確進行牛、羊和豬的發情鑑定。

• 規範完成對輸精器械的清洗和消毒。

• 規範完成各畜禽的輸精操作。

Project Guidance

　　Estrus identification is the most basic and key technology in animal reproduction. Learn the anatomy and function of the female reproduction organs, the growth of the follicles, estrous cycle, and the regulation of reproductive hormones, etc., so as to detect the time of estrus, determine proper time for insemination, standardize the insemination procedure, improve the rate of fertilization and conception.

Learning Objectives

>>> Knowledge Objectives

• To understand the anatomy and main function of female reproduction organs.

• To master the concept, classification and application of reproduction hormones.

• To master the physiology regulation of estrus.

• To know the growth of the follicles and the ovulation time.

• To understand the regulatory effects of reproductive hormones on estrus cycle.

>>> Skill Objectives

• To master the technology of estrus detection in cows, ewes and sows.

• To standardize the cleaning and disinfection of insemination devices.

• To standardize the insemination of livestock and poultry.

相關知識

一、雌性動物生殖器官的結構與功能

雌性動物的生殖器官主要由卵巢、輸卵管、子宮、陰道、陰唇、陰蒂等組成。幾種常見母畜生殖器官見圖3-1。母禽的生殖器官僅包括卵巢和輸卵管兩部分,家禽只有左側生殖器官,右側生殖器官在孵化的第7～9天就停止發育,到孵出時已退化,僅留殘跡。

Relevant Knowledge

1 Structure and functions of female reproductive organs

The female reproductive organs consist of the ovary, oviduct, uterus, vagina, vulva and clitoris (Figure 3-1). The hen has a pair of ovaries and oviducts. On the 7th to 9th day of hatching, the right-side ovary and oviduct will stop developing until only a vestige remains in the end. Therefore, the sexually mature hen only has a well-developed ovary and oviduct on the left side.

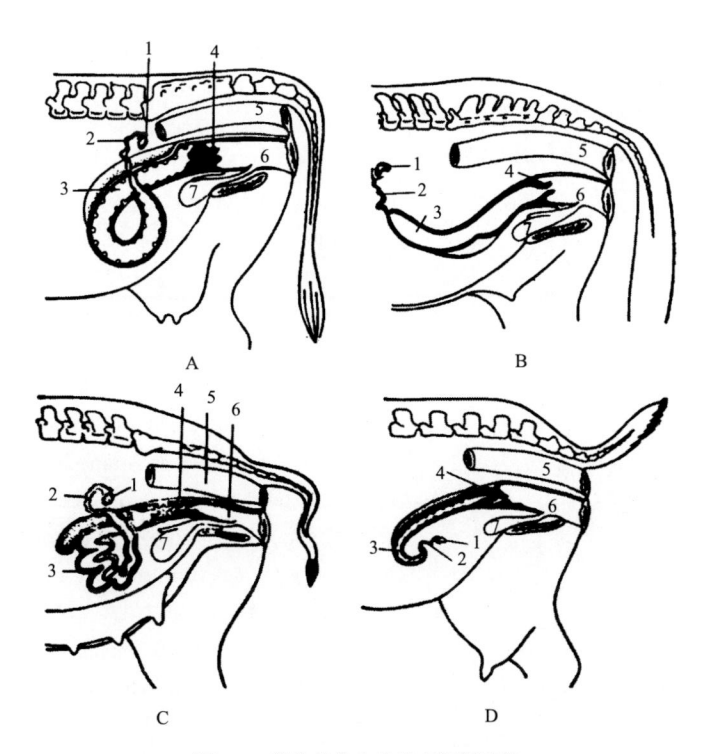

圖3-1　雌性動物生殖器官解剖示意

Figure 3-1　Reproductive organs of the female farm mammal

A. 牛 cow　B. 馬 mare　C. 豬 sow　D. 羊 ewe

1. 卵巢　2. 輸卵管　3. 子宮角　4. 子宮頸　5. 直腸　6. 陰道　7. 膀胱

1. ovary　2. oviduct　3. uterus horn　4. cervix　5. rectum　6. vagina　7. urinary bladder

(一) 卵巢

1. 形態和位置

卵巢是雌性動物產生卵子和分泌激素的重要結構，左右各一，附著在卵巢繫膜上。各種母畜的卵巢形狀如圖 3-2 所示。

牛的卵巢為扁橢圓形，位於子宮角尖端的兩側。羊的卵巢比牛的圓而小，位置與牛相同。初生仔豬的卵巢呈腎形，淡紅色，位於薦骨岬兩旁稍後方或在骨盆腔前口兩側的上部；當接近性成熟時（4～5月齡），因卵泡發育而呈桑葚形，性成熟後卵泡突出於卵巢表面，此時卵巢似串狀葡萄。馬的卵巢呈蠶豆形，表面光滑，附著緣寬大，游離緣有一凹陷，稱為排卵窩，這是馬屬動物特有的一個結構，其卵子只從此排出。

禽卵巢以短的繫膜附著於左腎前部。幼禽為扁平形，表面略呈顆粒狀。產蛋期卵巢如葡萄狀，有肉眼可見卵泡 1 000～1 500 個，成熟卵泡 4～5 個。停產期卵巢回縮，到下一個產蛋期又開始生長。

2. 機能

（1）卵泡發育和排卵。卵巢皮質部表層分布著許多原始卵泡，原始卵泡經過初級卵泡、次級卵泡、三級卵泡、成熟卵泡等幾個發育階段，部分卵泡最終發育為成熟卵子，由卵巢排出，並在原卵泡腔處形成黃體。

（2）分泌雌激素和孕激素。卵泡發

1.1 Ovary

1.1.1 Morphology and location

Ovaries produce ovum and secrete hormones. It adheres to the mesovarium of ovary both with one on the left side and the other on the right. Different breeds of females may have different shapes of ovaries(Figure 3-2).

The ovaries of cow mainly shape like flat ellipse, located on both sides of the tip of uterine horns. The ovaries of ewe are smaller to those of cow, and they are in the same position as cow. The ovaries of piglets are kidney-shaped and reddish in color. They are located at the posterior sides of the sacrum or at the upper sides of the anterior opening of the pelvic cavity. When becoming mature(about 4-5 months old), the follicles develop in the shape of mulberries, which extend out of the surface of the ovaries after sexual maturation, resembling strings of grapes. The ovaries of mare are broad bean-shaped with smooth surface, wide attach-ment margin and a depression (ovulation fossa) in the free margin. This is a unique structure of mare, and its ovum discharged from here.

The ovaries of poultry attach to the anterior part of the left kidney with a short mesangium. The poultry's ovary is flat with a slightly granular surface. During laying period, ovaries are grape-like, with 1 000-1 500 follicles and 4-5 mature follicles visible to the naked eye. The ovaries shrink during non-laying period and begin to grow again in the next laying period.

1.1.2 Function

(1) Follicular development and ovulation. There are many primordial follicles in the surface of ovarian cortex. It passes through several stages of development, such as primary follicle, secondary follicle, tertiary follicle and mature follicle. Some follicles eventually develop into mature eggs. After the ovum is discharged from the mature follicle, and the cavity of follicle changes into corpus luteum.

(2) Secretion of estrogen and progesterone. During

育過程中，卵泡內膜細胞可分泌雌激素，雌激素是導致母畜發情的直接因素。黃體可分泌孕激素，孕激素是維持母畜妊娠所必需的激素之一。

follicular development, follicular endometrial cells can secrete estrogen, which is the direct factor leading to estrus in female animals. The corpus luteum can secrete progesterone, which is one of the hormones to maintain the pregnancy.

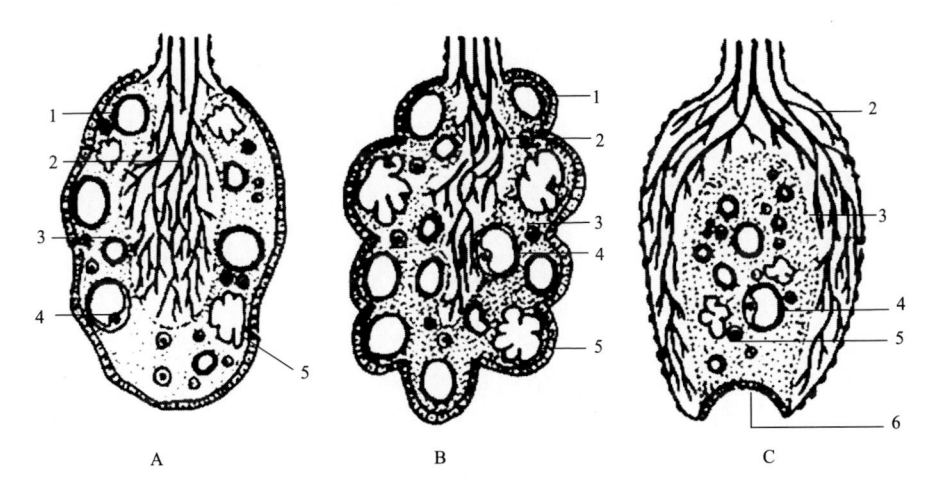

圖 3-2　母畜的卵巢形狀

Figure 3-2　Different shapes of female livestock's ovaries

A. 牛 cow　B. 豬 sow　C. 馬 mare

1. 生殖上皮　2. 髓質部　3. 皮質部　4. 卵泡　5. 黃體　6. 生殖上皮（排卵窩）

1. reproductive epithelium　2. medulla　3. cortical　4. follicle　5. corpus luteum　6. reproductive epithelium（ovulation fossa）

（二）輸卵管

1. 形態和位置

輸卵管位於卵巢和子宮角之間，形狀彎曲，長 15～30cm。輸卵管可分為漏斗部、壺腹部和峽部 3 個部分。

2. 機能

由卵巢排出的卵子先被輸卵管傘部接納（圖 3-3），然後向壺腹部運送。同時進入輸卵管的精子獲得受精能力，在壺腹部完成受精作用，並進行早期卵裂。

1.2　Oviduct

1.2.1　Morphology and location

The oviduct is located between the ovary and the uterine horn, with a curved shape and a length of 15-30 cm. It consists of infundibulum, ampulla and isthmus.

1.2.2　Function

The ovum discharged from the ovaries is caught by the umbrella of oviduct and then transported to ampulla (Figure 3-3). At the same time, the sperm entering the oviduct acquires fertilization ability and completes fertilization, then the early cleavage begins.

圖 3-3　輸卵管傘部接納卵子
Figure 3-3　Ovum caught by the umbrella of oviduct

（三）子宮

1. 形態和位置

子宮大部分位於腹腔，小部分位於骨盆腔，背側為直腸，腹側為膀胱。多數動物的子宮都由子宮角、子宮體和子宮頸三部分組成。牛、羊的子宮角彎曲如綿羊角（圖 3-4A、圖 3-4B），兩角基部之間有縱隔將兩子宮角分開，稱為對分子宮（也稱雙間子宮）。豬的子宮有兩個長而彎曲的子宮角（圖 3-4C、圖 3-4D），兩角基部之間的縱隔不明顯，為雙角子宮。

2. 機能

子宮可以篩選、儲存和運送精子，促進精子獲能；有利於孕體的附植、妊娠和分娩；調節卵巢黃體功能，引起發情等。

1.3　Uterus

1.3.1　Morphology and location

Uterus is mostly located in the abdominal cavity, with a small part in the pelvic cavity. The dorsal side is rectum and the ventral side is bladder. The uterus of most animals consists of uterus horn, uterus body and uterus cervix. The uterus horns of cow and ewe are crooked like the horns of sheep (Figure 3-4A and Figure 3-4B). There is a mediastinum at the uterus corner base. This kind of uterus is also called bifurcated uterus. The pig's uterus has two long, curved horns. The mediastinum of sow is not obvious between the two uterus corner bases, and it is called bicornute uterus(Figure 3-4C and Figure 3-4D).

1.3.2　Function

The uterus has the following functions: ①screening, storing and transporting the sperm, promoting sperm capacitation. ② beneficial to conceptus transplants, pregnancy and delivery. ③ regulating ovarian luteal function, leading to estrus and other functions.

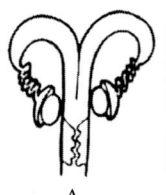

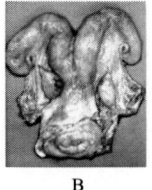

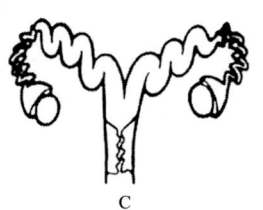

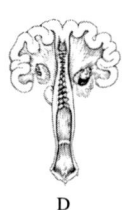

圖 3-4　子宮
Figure 3-4　Uterus
A、B. 牛 cow　C、D. 豬 sow

1.4 The vagina

The vagina serves as the female organ of copulation at mating and as the birth canal at parturition. It connects with the cervix in the anterior and extends to urethral reproductive vestibule in the posterior. The vagina serves the dual role of a passageway for the reproductive and urinary systems. Its mucosal surface changes during the estrous cycle from very moist when the animal is ready for mating to almost dry, even sticky. Its biochemical and microbial environment can protect the upper genital tract from microbial invasion.

2 Reproductive hormones

Hormone is a chemical substance produced in the body that controls and regulates the activity of certain cells or organs. The hormone that directly acts on reproductive activities and regulates reproductive function and reproductive process is called reproductive hormone(Table 3-1).

Reproductive hormones are closely related to the physiological activities of female animals, such as ovum formulation, follicles development, ovulation, estrus cycle changes, fertilization, pregnancy, parturition and lactation. Reproductive hormones are not only widely used in various reproductive technologies, but also used as a kind of regulator, which is useful to regulate animal estrus, prevention and control of reproductive disorders. It shows an effective result in the treatment of infertility especially.

Table 3-1 Name, source and function of main reproductive hormones

名稱 Name	簡稱 Abbreviation	來源 Source	生理功能 Function
促性腺激素釋放激素 Gonadotropin releasing hormone	GnRH	下視丘 Hypothalamus	促進 LH 和 FSH 釋放 Stimulate the release of LH and FSH

(續)

名稱 Name	簡稱 Abbreviation	來源 Source	生理功能 Function
催產素 Oxytocin	OXT	下視丘 Hypothalamus	促進子宮收縮和排乳 Promote uterine contraction and lactation
促卵泡素 Follicle-stimulating hormone	FSH	垂體前葉 Anterior pituitary	促進卵泡發育和精子發生 Promote follicular development and spermatogenesis
促黃體素 Luteinizing hormone	LH	垂體前葉 Anterior pituitary	促進排卵，形成黃體，促進孕酮分泌 Promote ovulation, corpus luteum formation and progesterone secretion
促乳素 Prolactin	PRL	垂體前葉 Anterior pituitary	促進泌乳，增強母性行為 Promote lactation, enhance maternal behavior
孕馬血清促性腺激素 Pregnant mare's gonadotrophin	PMSG	馬胎盤 Horse placenta	具有 FSH 和 LH 雙重活性，以 FSH 為主 Double bio-activities of FSH (main) and LH
人絨毛膜促性腺激素 Human chorionic gonadotrophin	HCG	靈長類胎盤絨毛膜 Primate placental chorion	與 LH 相似 Similar to LH
雌激素 Estrogen	E_2	卵巢、胎盤 Ovary and placenta	促進動物發情行為，維持第二性徵，刺激生殖器官的發育 Promote the estrus behavior, maintain the secondary sexual characteristics and stimulate the development of reproductive organs
孕激素 Progesterone	P_4	卵巢（黃體）、胎盤 Corpus luteum and placenta	調節發情，維持妊娠，促進乳腺發育 Regulate estrus, maintain pregnancy and promote mammary development
雄激素 Androgen	A	睪丸間質細胞 Interstitial cells of testis	維持雄性動物第二性徵和性行為，促進精子發生 Maintain the secondary sexual characteristics and sexual behavior of male animals, promote spermatogenesis
鬆弛素 Relaxin	RLX	卵巢、胎盤 Ovary and placenta	促使子宮頸擴張、骨盆韌帶鬆弛 Promote cervical dilatation and pelvic ligament relaxation
前列腺素 Prostaglandin	PG	子宮（內膜） Uterus (endometrium)	溶解黃體，促進子宮收縮 Regress corpus luteum, promote uterine contraction
外激素 Pheromone	PHE	外分泌腺 Exocrine glands	影響性行為和性活動 Influence sexual behavior

三、雌性動物的性機能發育

雌性動物的性機能發育是一個由發生、發展直至衰退停止的過程,一般分為初情期、性成熟期、體成熟期及繁殖機能停止期。為了指導生產,還涉及初配期。

1. 初情期

初情期是指雌性動物第一次出現發情現象和排卵的時期。到達初情期體重為成年體重的 30%～40%,雖有發情表現,但不完全,發情週期也往往不正常,其生殖器官仍在繼續生長發育中,所產生的卵子品質較差。此階段配種雖有受精的可能性,但不宜配種。

2. 性成熟期

初情期後,隨著年齡的增長,生殖器官進一步發育成熟,發情排卵活動已趨正常,具備了正常繁殖後代的能力,此時稱為性成熟。性成熟期時期,動物體重占成年體重的 50%～60%,身體發育尚未成熟,故一般不宜配種。

3. 體成熟期

一般當雌性動物體重達到成年體重的 70% 時就可以配種。雌性動物在初配後受胎,身體仍未完全發育成熟,經過一段時間後才能達到體成熟。

4. 繁殖機能停止期

雌性動物經過多年的繁殖活動,生殖器官逐漸老化,繁殖機能逐漸衰退,甚至喪失繁殖能力。在畜牧生產中,一般在母畜繁殖機能停止之前,只要生產效益明顯下降,就對其進行淘汰。

3 Sexual development of female animal

Sexual development of female animal is a process including the puberty stage, sexual maturity stage, adult stage and reproductive cessation stage. In practical production, we also consider the initial mating stage.

3.1 Puberty stage

Puberty stage refers to the period when the female animal appears estrus and ovulation for the first time. The body weight is about 30%-40% of the adutt's body weight at this stage. The estrous behavior is not complete, and the estrus cycle is often abnormal. Their reproductive organs are still growing and developing, and the quality of ovum is not good. In conclusion, female is not suitable for mating at this stage.

3.2 Sexual maturity stage

After the puberty stage, as the female grows up, reproductive organs are getting mature, estrus and ovulation activities are more regular. At this stage, the female has the ability of reproduction. During sexual maturity stage, the body weight accounts for 50%-60% of the adult's body weight, which means the body has not mature yet, so it is still not suitable for mating.

3.3 Adult stage

When the body weight of female animal accounts for 70% of the adult's, we can use it for breeding. After initial mating, the body is still not fully mature. It takes some time for them to reach full body maturity.

3.4 Reproductive cessation stage

After years of reproduction, the ageing organs of female animals' reproductive system are gradually losing their reproductive ability. In animal husbandry production, we usually eliminate those female animals as long as their production efficiency obviously decreases. This usually happens before the reproductive cessation stage.

四、卵子的發生與卵泡發育

(一) 卵子的發生

1. 卵原細胞的增殖

動物在胚胎期性別分化後，雌性胎兒的原始生殖細胞便分化為卵原細胞。卵原細胞通過多次有絲分裂的方式增殖成許多初級卵母細胞。

2. 卵母細胞的生長

初級卵母細胞的生長是伴隨卵泡的生長而實現的，卵泡細胞為卵母細胞的生長提供營養。初級卵母細胞階段的主要特點：卵黃顆粒增多，卵母細胞體積增大，出現透明帶，卵母細胞周圍的卵泡細胞通過有絲分裂增殖，由扁平形單層變為立方形多層。初情期到來之前，卵母細胞生長發育處於停滯狀態。

3. 卵母細胞的成熟

卵母細胞的成熟需經過兩次成熟分裂。卵泡中的卵母細胞是一個初級卵母細胞，在排卵前不久完成第一次成熟分裂。大多數動物在排卵時，卵子尚未完成成熟分裂。牛、綿羊和豬的卵子，在排卵時只是完成第一次成熟分裂，即卵泡成熟破裂時，放出次級卵母細胞和一個極體，排卵後次級卵母細胞開始第二次成熟分裂，直到精子進入透明帶，卵母細胞被啟動後，放出第二極體，這時才完成第二次成熟分裂。一個初級卵母細胞經分裂最終發育形成1個卵子和3個極體。

(二) 卵子的形態和結構

1. 卵子的形態和大小

4 Oogenesis and follicular development

4.1 Oogenesis

4.1.1 Oogonia proliferation

After sex differentiation in the embryonic stage, the primordial germ cells of the female fetus differentiate into oogonia. Through multiple mitosis, oogonia proliferate into many primary oocytes.

4.1.2 Oocyte growth

The growth of primary oocyte is accompanied by the growth of follicle, which provides the nutrition for the growth of oocyte. There are some main characteristics of primary oocyte stage, including: the number of yolk granules increases, the volume of oocyte increases, zona pellucida appears, and the follicular cells surrounding oocyte proliferate through mitosis, changing from flat monolayer to cubic multilayer. Before the puberty, the development of oocyte is arrested at primary oocyte stage.

4.1.3 Oocyte maturation

The process of oocyte maturation needs two maturation divisions. The oocyte in the follicle is a primary oocyte with completion of the first maturation division shortly prior to ovulation. The ovum has not completed maturation division yet while most animals ovulating. The ovum from cow, ewe and sow only has completed the first maturation division when ovulation occurs. In details, one secondary oocyte and one first polar body are released at the time of the follicle maturation and rupture. After ovulation, the secondary oocyte begins the second maturation division until the sperm enters the zona pellucida. Until the sperm enters the zona pellucida and the oocyte is activated, the second polar body is released, and then the second maturation division is completed. Eventually, one primary oocyte develops to form one ovum and three polar bodies.

4.2 Morphology and structure of ovum

4.2.1 Shape and size of ovum

哺乳動物的卵子為圓球形，卵子較一般細胞含有更多的細胞質，大多數哺乳動物的卵子，直徑為 $70\sim140\mu m$。

2. 卵子的結構

卵子的主要結構包括放射冠、透明帶、卵黃膜及卵黃等部分（圖3-5）。

The ovum of mammal is spherical with $70\text{-}140\mu m$ in diameter, and contains more cytoplasm than ordinary cells.

4.2.2　Ovum structure

The main structure of the ovum includes the corona radiate, the zona pellucida, the vitelline membrane and the ooplasm, et al.（Figure 3-5）

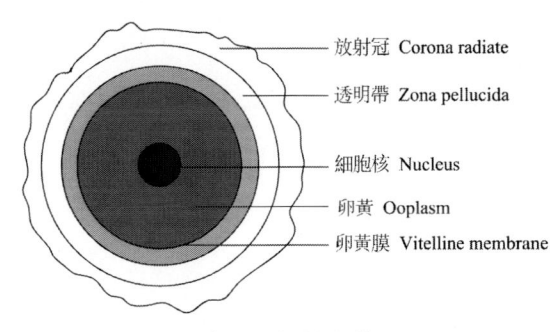

放射冠　Corona radiate
透明帶　Zona pellucida
細胞核　Nucleus
卵黃　Ooplasm
卵黃膜　Vitelline membrane

圖 3-5　卵子的結構
Figure 3-5　Structure of the ovum

（1）放射冠。卵子周圍緻密的顆粒細胞呈放射狀排列，故名放射冠。放射冠位於卵子最外層，為卵母細胞提供養分並進行物質交換。

（2）透明帶。為一均質的蛋白質半透明膜，一般認為它由卵泡細胞和卵母細胞形成的細胞間質組成。其作用是保護卵子，在受精時發生透明帶反應，防止多個精子進入，使受精正常進行。同時，透明帶存在與精子特異性結合的位點，使異種動物的精子不能和透明帶結合。

（3）卵黃膜。是卵母細胞的皮質分泌物，其作用主要是保護卵母細胞完成正常的生命活動，以及在受精過程中發生卵黃膜封閉作用，阻止多精子受精。

（4）卵黃。排卵時卵黃較接近透

（1）Corona radiate. The dense cumulus cells surround the ovum radially, which is called corona radiate. The corona radiate is located at the outermost layer of the ovum, providing nutrients to the oocyte and exchanging substances.

（2）Zona pellucida. The zona pellucida is a homogeneous translucent membrane, made of intercellular substances formed by follicular cells and oocytes generally. It protects ovum, causes the acrosome reaction during fertilization, prevents post-fertilization polyspermy. The specific binding region of sperm appears in the zona pellucid, which prevents xenogeneic species' sperm from binding with the zona pellucid.

（3）Vitelline membrane. It is the secretion of the oocyte cortex, which can protect the oocytes to complete the normal life events and undergo the vitelline membrane reaction to block polyspermy during fertilization.

（4）Ooplasm. It is closer to the zona pellucida at

帶，受精後卵黃收縮，透明帶和卵黃膜之間的卵黃周隙擴大，排出的極體即在周隙之中。此外，卵黃為卵子和早期胚胎發育提供營養物質。

（5）卵核。位於卵黃內，雌性動物的主要遺傳物質就分布在卵核內。

（三）卵泡的生長發育

卵泡是包裹卵母細胞或卵子的特殊結構。在胚胎期，雌性動物就已形成大量原始卵泡儲存於卵巢皮質部，卵泡的生長貫穿於胚胎期、幼齡期和整個生育期。卵泡發育從形態上可分為原始卵泡、初級卵泡、次級卵泡、三級卵泡和成熟卵泡5個階段（圖3-6）。

1. 原始卵泡

原始卵泡位於卵巢皮質外周，是體積最小的卵泡，其核心為一卵母細胞，周圍為一單層扁平狀的卵泡上皮細胞，沒有卵泡膜和卵泡腔。在胎兒期已有大量原始卵泡作為儲備，除少數發育成熟外，其他均在發育過程中閉鎖、退化而死亡。

2. 初級卵泡

初級卵泡由原始卵泡發育而成，排列在卵巢皮質外圍，是由卵母細胞和周圍的一層立方形卵泡細胞組成，卵泡膜尚未形成，也無卵泡腔。

3. 次級卵泡

在生長發育過程中，初級卵泡移向

the time of ovulation. After fertilization, the vitelline shrinks, and the perivitelline space between the zona pellucida and vitelline membrane becomes expanded, where the released polar bodies appear. In addition, yolk provides nutrients for ovum and early embryo development.

(5) Nucleus. The mainly genetic material is carried in the nucleus, which is located in the vitelline.

4.3 Growth and development of follicular

Follicle is a special structure around oocyte or ovum. In the embryonic stage, a large number of primordial follicles are located in the ovarian cortex, and the process of follicular growth occurs throughout the embryonic stage, juvenile stage and the child bearing period. Follicular development in morphology can be divided into five stages: primordial follicle, primary follicle, secondary follicle, tertiary follicle and mature follicle(Figure 3-6).

4.3.1 Primordial follicle

It is the smallest follicle, located in the outer periphery of the ovarian cortex. The core of primitive follicles is an oocyte surrounded by a single layer of flat follicular epithelial cells. It has no follicular membrane or follicular cavity. There are a large number of primordial follicles as reserves in the fetal period. Except for a few mature, the others are locked, degenerated and die during development.

4.3.2 Primary follicle

It develops from the primordial follicle, arranging on the periphery of the ovarian cortex, and is composed of the oocyte surrounded by a layer of cuboid follicular cells. In primary follicle, the follicular membrane has not yet formed and the follicular cavity is absent.

4.3.3 Secondary follicle

During growth and development, the primary fol-

卵巢皮質的中央，卵泡上皮細胞增殖形成多層立方形細胞，細胞體積變小，稱為顆粒細胞。隨著卵泡的生長，卵泡細胞分泌的液體聚集在卵黃膜與卵泡細胞（或放射冠細胞）之間形成透明帶。此時尚未形成卵泡腔。

4. 三級卵泡

三級卵泡由次級卵泡發育而成。在這一時期中，卵泡細胞分泌的液體，使卵泡細胞之間分離，並使卵母細胞之間的間隙增大，形成不規則的腔隙，稱為卵泡腔。隨著卵泡液的增多，卵泡腔也逐漸擴大，卵母細胞被擠向一邊，並被包裹在一團顆粒細胞中，形成半島突出於卵泡腔中，稱為卵丘。其餘的顆粒細胞緊貼於卵泡腔的周圍，形成顆粒層。

5. 成熟卵泡

三級卵泡繼續生長，卵泡壁變薄，卵泡液增多，卵泡腔增大，卵泡擴展到整個卵巢的皮質部而突出於卵巢的表面。各種動物成熟卵泡的大小差異很大，牛的直徑為 10～14mm，豬為 8～12mm，綿羊為 5～10mm，山羊為 7～10mm。多胎動物在一個發情週期裡，可有數個至數十個原始卵泡同時發育到成熟卵泡，而單胎動物一般只有一個卵泡發育成熟並排卵。

（四）卵泡的閉鎖與退化

動物出生前，卵巢上就有很多原始卵泡，但只有少數卵泡能夠發育成熟並排卵，絕大多數卵泡發生閉鎖和退化。越是年輕的動物，卵泡閉鎖的發生越嚴

licle moves to the center of the ovarian cortex, and the follicular epithelial cells proliferate to form a several layers of cuboidal cells with smaller volume. As the follicles grow, the fluid secreted by the follicular cells accumulates between the vitelline membrane and the follicular cells (or the corona radiate cells) to form zona pellucida. The follicular cavity has not yet formed.

4.3.4 Tertiary follicle

It develops from the secondary follicles. In this period, the fluid secreted by the follicular cells separates the follicular cells from each other, expands the space between the oocytes, and eventually forms an irregular cavity, which is called the follicular cavity. As the follicular fluid increases, the follicular cavity gradually expands. The oocyte squeezed to one side is surrounded by a mass of granulosa cells, forming a peninsula into the follicular cavity, which is called the cumulus. The other granulosa cells are closely attached to the periphery of the follicular cavity to form the granular layer.

4.3.5 Mature follicle

The tertiary follicles continue to grow with follicular wall getting thinner, the follicular fluid increasing, and follicular cavity expanding. The follicles extend to the cortex of the whole ovary and protrude from the surface of the ovary. The sizes of the mature follicles from different animals vary greatly. The diameter of mature follicles of the cattle is 10-14 mm, the pig is 8-12 mm, the sheep is 5-10 mm, and the goat is 7-10 mm. In an estrus cycle, multiple primordial follicles of multiparous animals can develop into mature follicles at the same time, while only one follicle of the uniparous animal can mature and ovulate generally.

4.4 Follicular atresia and degeneration

Before birth, many primordial follicles exist in the ovary, but only a few of them can mature and ovulate, and most of them degenerated. The younger the animal is, the more severe the occurrence of follicular atresia is. The cause of follicular atresia may be attributed to

重。卵泡發生閉鎖的原因可能與垂體前葉FSH的分泌不足或卵泡對FSH的反應性降低有關。

insufficient secretion of FSH in the anterior pituitary or decreased reactivity of the follicles to FSH.

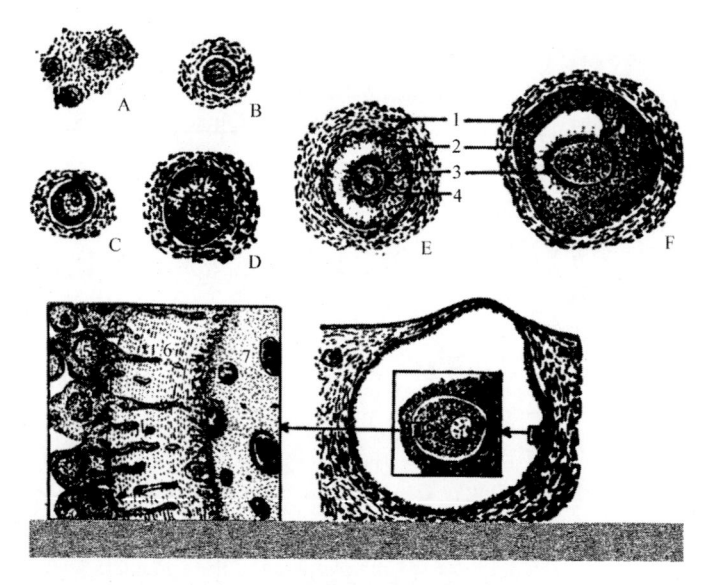

圖 3-6　哺乳動物卵泡發育模式圖

Figure 3-6　Pattern of follicular development in mammals

A. 原始卵泡　B. 初級卵泡　C. 次級卵泡　D. 三級卵泡　E. 出現新月形腔隙的三級卵泡
F. 出現卵丘的三級卵泡　G. 卵黃膜的微絨毛部分伸向透明帶　H. 成熟卵泡

A. primordial follicle　B. primary follicle　C. secondary follicle　D. tertiary follicle
E. tertiary follicle with crescent follicular cavity　F. tertiary follicle with the cumulus
G. microvilli of yolk membrane extending to zona pellucida　H. mature follicle

1. 卵泡外膜　2. 顆粒層　3. 透明帶　4. 卵丘　5. 顆粒層細胞　6. 透明帶　7. 卵黃
1. follicular adventitia　2. granular layer　3. zona pellucida　4. cumulus　5. granulosa cell
6. zona pellucida　7. ooplasm

（五）排卵

成熟卵泡破裂、釋放卵子的過程，稱為排卵。生長發育到一定階段，卵泡成熟破裂，隨著卵泡液的流出，卵子與卵丘脫離而被排出卵巢外，由輸卵管傘部接納。

1. 排卵的類型

（1）自發性排卵。卵巢上的成熟卵泡自行破裂排卵並自動形成黃體。這種

4.5　Ovulation

The mature follicles rupture, thus freeing the ovum in a process called ovulation. At a certain stage of growth and development, the follicle matures and ruptures, then frees as the follicular fluid drains. The ovum separated from the cumulus is discharged out from the ovary, and enters the fimbriae of the oviduct.

4.5.1　Types of the ovulation

(1) Spontaneous ovulation. The mature follicle on the ovary ruptures and spontaneously forms the corpus

類型又有兩種情況：一是發情週期中黃體的功能可以維持一定時期，且具有功能性，如牛、豬、馬、羊等屬於此種類型；二是除非交配（刺激），否則形成的黃體是非功能性的，即不分泌孕酮，鼠屬於這種類型。

（2）誘發性排卵。又稱為刺激性排卵，即卵泡發育成熟後，必須通過交配或其他途徑使子宮頸受到機械性刺激後才能排卵，並形成功能性黃體。兔、駱駝、貓等屬於誘發性排卵動物，牠們在發情季節中，卵泡有規律地陸續成熟和退化，如果交配（刺激），隨時都有成熟的卵泡排卵。

2. 排卵時間

排卵是成熟卵泡在 LH 峰作用下產生的，各種動物的排卵時間因動物種類、品種、個體、年齡、營養狀況及環境條件等不同而異。常見家畜的排卵時間如下：牛在發情結束後 8～12h，豬在發情開始後 16～48h，羊在發情結束時，兔在交配刺激後 6～12h。

3. 排卵數目

牛、馬等大家畜一般為 1 枚，個別的可排 2 枚；綿羊 1～3 枚；山羊 1～5 枚；豬 10～25 枚；兔 5～15 枚。

（六）黃體型成與退化

成熟卵泡排卵後形成黃體，黃體分泌孕酮作用於生殖道，使之向妊娠方向變化。如未受精，一段時間後黃體退化，開始下一次的卵泡發育與排卵。

luteum. Firstly, the function of the corpus luteum can be maintained in a certain period of estrus cycle, such as cows, sows, mares, ewes, etc. Secondly, unless mating (stimulation), the formed corpus luteum would lose its function, which means the progesterone is not secreted by the corpus luteum, such as the case in rats.

(2) Induced ovulation. It is also known as stimulating ovulation. After follicles maturation, the cervix must be mechanically stimulated by mating or other ways to ovulate and form functional corpus luteum. Induced ovulation occours in animals such as rabbits, camels, cats and so on. During the estrus season, their follicles mature and degenerate regularly and gradually. However, if mating (stimulation) occurs, mature follicles will ovulate at any time.

4.5.2 The time of ovulation

Ovulation occurs under the LH peak stimulated by mature follicles. The time of ovulation varies according to animal species, breed, individual, age, nutritional status and environmental conditions. The time of ovulation is 8-12 hours after end of estrus in cows, 16-48 hours after onset of estrus in sows, near the end of estrus in ewes, and 6-12 hours after the stimulation of mating in rabbits.

4.5.3 The number of ovulation

The number of ovulation is one generally, occasionally two in large domestic animals such as cows and mares, 1-3 in ewes, 1-5 in goats, 10-25 in sows, and 5-15 in rabbits.

4.6 The formation and degeneration of the corpus luteum

After ovulation of mature follicles, it forms the corpus luteum, which secretes progesterone and stimulates the reproductive tract to prepare for pregnancy. In non-fertilized animals, the corpus luteum gradually degenerates and starts the next follicular development and ovulation.

成熟卵泡破裂排卵後，卵泡腔產生負壓，卵泡膜血管破裂流血，並充溢於卵泡腔內形成血凝塊，稱為紅體。此後顆粒細胞在 LH 作用下增生變大，並吸取類脂質而變成黃體細胞。同時卵泡內膜血管增生分布於黃色細胞團中，卵泡膜的部分細胞也進入黃色細胞團，共同構成了黃體。黃體在排卵後 7～10d（牛、羊、豬）或 14d（馬）發育至最大體積。

　　黃體是一種暫時性的分泌組織，主要作用是分泌孕酮。在發情週期中，如果雌性動物沒有妊娠，所形成的黃體在黃體期末退化，這種黃體稱為週期性黃體。週期性黃體通常在排卵後維持一定時間才退化，退化時間牛為 14～15d，羊為 12～14d，豬為 13d，馬為 17d。如果雌性動物妊娠，則黃體存在的時間長，體積也增大，這種黃體稱為妊娠黃體。妊娠黃體分泌孕酮以維持妊娠需要，直至妊娠結束時才退化。但馬、驢的妊娠黃體退化較早，一般在妊娠後 160d 即開始退化，以後靠胎盤分泌孕酮維持妊娠。

五、發情生理

　　雌性動物生長發育到一定年齡後，受下視丘-垂體-卵巢軸調控，每隔一定時間，卵巢上就有卵泡發育，逐漸成熟而排卵，此時雌性動物的精神狀態和生

After the rupture and ovulation of the mature follicle, the follicular cavity produces the negative pressure, with the blood vessels on the follicular membrane rupture and blood, and the cavity of the ruptured follicle fills with a blood clot, which is called corpus hemorrhagic. Thereafter, the granulosa cells proliferate under the action of LH, and absorb lipoid to develop into the luteal cells. At the same time, the vascular proliferation on follicular endometriosis distribute into the yellow cell mass, and some cells of the follicular theca also enter the yellow cell mass, which join together to form the corpus luteum. The corpus luteum develops to the maximum volume in 7-10 days(cow, ewe and sow) or 14 days(mare) after ovulation.

　　The corpus luteum is a temporary secretory tissue. Its main function is to secrete progesterone. During an estrous cycle, if the female animal does not have a pregnancy, the formed corpus luteum degenerates at the end of the luteal phase, which is called periodic corpus luteum. The periodic corpus luteum usually lasts for a time after ovulation, and then degenerate. The time of the regression is 14-15 days in cows, 12-14 days in ewes, 13 days in sows, 17 days in mares. If the female animal is pregnant, the corpus luteum lasts for a long time and enlarges in volume, this cropus luteum is called pregnant corpus luteum. The pregnant corpus luteum secretes progesterone to maintain pregnancy, and it does not degenerate until the end of pregnancy. However, the corpus luteum degenerates earlier in horse and donkey. It usually begins to degenerate in 160 days after pregnancy, afterwards the pregnancy is maintained by the progesterone secreted by the placenta.

5　Estrus physiology

　　When the female animal grows and develops to a certain age, at the regular intervals, the follicles on the ovary matures gradually to ovulate, and this process is regulated by the hypothalamic-pituitary-ovarian axis. At this time, the mental state, reproductive organs and

殖器官及行為都發生一系列變化，如興奮不安、食慾減退、外陰腫脹、陰道黏液增多，有強烈的求偶行為。

（一）發情週期

雌性動物在初情期以後，表現出週而復始的性活動週期。即在生理或非妊娠條件下，母畜每間隔一定時期均會出現一次發情，通常把一次發情開始至下次發情開始，或一次發情結束至下次發情結束所間隔的時期，稱為發情週期。牛、豬、山羊的發情週期平均為 21d，綿羊為 16～17d。根據機體所發生的一系列生理變化，發情週期多採用四期分法和二期分法（圖 3-7）。

behaviors of the female animals all undergo a series of changes, such as excitement and unrest, appetite loss, vulvar swelling and mucus increasing and display strongly courtship behaviors.

5.1 Estrus cycle

Females show a cycle of sexual activity after puberty. Under physiological or non-pregnant conditions, an estrus occurs periodically at regular time intervals in females. Usually, the interval period, from the onset of an estrus to the onset of next estrus, or from the end of an estrus to the end of next estrus, is called an estrous cycle. The estrous cycle is 21 days on average in cows, sows and female goats, and 16-17 days in ewes. Based on a series of physiological changes, an estrous cycle is usually divided into various phases using four-phase or two-phase methods generally (Figure 3-7).

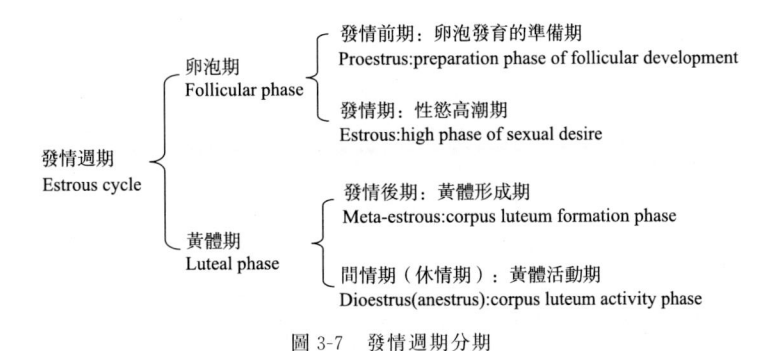

發情週期 Estrous cycle
- 卵泡期 Follicular phase
  - 發情前期：卵泡發育的準備期 Proestrus: preparation phase of follicular development
  - 發情期：性慾高潮期 Estrous: high phase of sexual desire
- 黃體期 Luteal phase
  - 發情後期：黃體形成期 Meta-estrous: corpus luteum formation phase
  - 間情期（休情期）：黃體活動期 Dioestrus(anestrus): corpus luteum activity phase

圖 3-7　發情週期分期
Figure 3-7　Phases of the estrous cycle

（二）發情持續期

發情持續期是指母畜從發情開始到發情結束所持續的時間，相當於發情週期中的發情期。各種母畜的發情持續期為：牛 1～2d，羊 1～1.5d，豬 2～3d，馬 4～7d。由於季節、飼養管理水平、年齡及個體條件的不同，母畜發情持續期的長短也有所差異。

5.2 Duration of estrus

The duration of estrus is the length of time from the onset of estrus to the end of estrus, equivalent to estrus phase in the estrus cycle. The duration of estrus in various females is as follows: 1-2 days in cows, 1-1.5 days in ewes, 2-3 days in sows, and 4-7 days in mares. The variable length of the duration of the estrous in various females depends on differences in season, feeding management, age and individual condi-

(三) 發情季節

雌性動物發情主要受神經內分泌調控，但也受外界環境條件的影響，季節變化是影響發情週期的重要環境因素。有些動物如馬、綿羊、犬等，一年中只在特定季節才表現出發情。動物的發情可分為季節性發情和全年發情兩種類型。季節性發情又分為季節性多次發情和季節性單次發情兩種。

季節性多次發情是指在發情季節有多個發情週期。如馬、綿羊在春季或秋季發情時，如果沒有配種或配種後未受胎，可出現多次發情。犬的發情季節為春、秋兩季，但在每個發情季節內只有一個發情週期。全年多次發情是指雌性動物在一年四季都可以出現發情並可配種，牛、豬等屬此類型。

(四) 異常發情

1. 安靜發情

安靜發情又稱為隱性發情，是指母畜發情時外部表現不明顯，但卵巢上有卵泡發育、成熟並排卵。常見於產後第一次發情以及營養不良的牛、馬和羊。雌激素分泌不足或體內缺乏孕酮均可引起安靜發情。

2. 短促發情

短促發情指母畜發情持續時間短，往往錯過配種時機。其原因可能是神經內分泌系統的功能失調，發育的卵泡很快成熟破裂排卵，縮短了發情期，也可能是由於卵泡突然停止發育或發育受阻而引起。

3. 斷續發情

斷續發情即母畜發情延續很長，且

tions.

5.3 Estrus season

The estrus in females is mainly regulated by neuroendocrine, and also affected by external environment. The seasonal changes are important factors affecting the estrus cycle. The estrus in some species, such as horses, sheep and dogs, occurs in a certain season. There are seasonal estrus and annual estrus. Seasonal estrus includes seasonal poly-estrus and seasonal mono-estrus.

The seasonal poly-estrus represents the poly-estrus cycles during the estrus season. For example, while a horse or sheep is estrus in spring or autumn, if they are not mated or un-pregnant, poly-estrus will occur. The estrus seasons in dogs are spring and autumn, but only one estrus cycle occurs in each estrus season. Poly-estrus yearly means that the females can be estrus and mated all year round, such as cows and sows.

5.4 Abnormal estrus

5.4.1 Silent estrus

Silent estrus, also known as recessive estrus, represents that the external signs are not obvious in the estrus female, but the follicles on the ovary mature, develop and ovulate normally. It is commonly observed on the first postpartum estrus or malnourished cows, mares and ewes. The silent estrus is caused by insufficient secretion of estrogen or loss of progesterone in vivo.

5.4.2 Short estrus

The short estrus represents a short duration of estrus in females. It is easy to miss the mating time. The reason may be the dysfunction of the neuroendocrine system, resulting that the developing follicles quickly mature, rupture and ovulate, thus shortening the estrus period. It also may be caused by sudden cessation of follicular development or obstruction of follicular development.

發情時斷時續。多見於早春或營養不良的母馬。其原因是卵泡交替發育的結果。

4. 持續發情

持續發情是慕雄狂的症狀之一，表現為持續強烈的發情行為。患慕雄狂的母牛，表現為極度不安，大聲哞叫，頻頻排尿，追逐爬跨其他母牛，產奶量下降，食慾減退，身體消瘦，外形往往具有雄性特徵，如頸部肌肉發達等。

5. 孕後發情

孕後發情又稱假發情，是指妊娠母畜仍有發情表現。母牛在妊娠最初3個月內，常有3%～5%的母牛發情，綿羊孕後發情可達30%。孕後發情發生的主要原因是激素分泌失調，即妊娠黃體分泌孕酮不足，而胎盤分泌雌激素過多所致，容易引起流產，稱之為「激素性流產」。生產上常常會因為誤配而造成流產，因此要認真區分。

（五）產後發情

產後發情是指母畜分娩後的第一次發情。乳牛一般在產後25～30d發情，多數表現為安靜發情，產後40～50d正常發情。母豬一般在分娩後3～6d內可出現發情，但不排卵。在仔豬斷奶後一週之內，80%左右的母豬出現第一次正常發情。母羊大多在產後2～3個月發情。母馬往往在產駒後6～12d發情，一般發情表現不太明顯，甚至無發情表現，但有卵泡發育且可排卵，若配種可受孕。

5.4.3 Intermittent estrus

This form of estrus in the females lasts for a long time, characterized by an intermittent process. It is common in early spring time and in malnourished mares, which can be attributed to the alternate development of follicles.

5.4.4 Continuous estrus

The continuous estrus is one of the symptoms of nymphomania, characterized by strongly persistent estrus behavior. The cows suffering from nymphomania are characterized by extremely uneasy, screaming loudly, frequent micturition, chasing and mounting other cows, reduction of milk production, loss of appetite, weight loss, and appearance of maleness involving muscular neck etc.

5.4.5 Estrus after pregnancy

It is also known as false estrus, which means that the pregnant females still have estrus behavior. The estrus often occurs in 3%-5% of cows in the first three months after pregnancy, and in 30% of ewes after pregnancy. The main cause of this phenomenon may be the imbalance of hormone secretion. Such as, insufficient progesterone secreted by pregnant corpus luteum and excessive estrogen secreted by the placenta, can cause abortion easily, which is called「hormonal abortion」. In practical production, it is necessary to distinguish the real estrus from the false ones after pregnancy to avoid abortion.

5.5 Postpartum estrus

It is the first estrus in females after delivery. The estrus generally occurs in cows in 25-30 days after delivery, mostly in the form of silent estrus, and it is seen as normal estrus after 40-50 days after delivery. Sows usually display estrus within 3-6 days after delivery without ovulation. About 80% of them reach the first estrus normally within one week after weaning of piglets. Estrus of ewes usually occurs within 2-3 months after delivery. Mares often reach the estrus within 6-12 days after delivery, with inconspicuous or no estrus be-

（六）乏情

乏情是指達到初情期的雌性動物長期不發情，卵巢無週期性的功能活動，而是處於相對靜止狀態。引起乏情的因素很多，有季節性、生理性和病理性等。

1. 季節性乏情

季節性乏情的動物在非繁殖季節，卵巢和生殖道處於靜止狀態。季節性乏情的時間因畜種、品種和環境而異。馬多為短日照的冬春季乏情，綿羊的乏情往往發生於長日照的夏季。

2. 生理性乏情

生理性乏情包括泌乳性乏情、妊娠期乏情和衰老性乏情等。豬是最常見的泌乳性乏情動物，泌乳期間發情和排卵受到抑制，一般在仔豬斷乳後才出現發情。妊娠期間由於卵巢上存在妊娠黃體，可以分泌孕酮而抑制發情。妊娠期乏情是保證胚胎正常發育的生理現象。動物因衰老使下視丘-垂體-性腺軸的功能減退，導致垂體促性腺激素分泌減少，或卵巢對這些激素的反應性降低。

3. 病理性乏情

營養不良、壓力反應、卵巢機能疾病（如持久黃體、黃體囊腫等）均會抑制發情。

havior, however the follicular development and ovulation still occur, and they can pregnant after mating.

5.6 Anestrus

Being anestrus means that the females reaching puberty can not begin the process of the estrus for a long time, and the ovary is in a relatively static state. There are many factors causing anestrus, such as seasonality, physiology and pathology reasons etc.

5.6.1 Seasonal anestrus

The ovary and reproductive tract of the seasonal anestrus animals are at rest in the non-breeding season. The time of seasonal anestrus varies depending on species, breed and environment. For example, the anestrus in mares occur in winter and spring with short-daylight, while the anestrus in ewes often occur in summer with long-daylight.

5.6.2 Physiological anestrus

Physiological anestrus includes lactational anestrus, pregnant anestrus and aging anestrus. Sows are the most common lactational anestrus animals. During lactation, estrus and ovulation are inhibited. Estrus usually occurs after weaning. The females can secret progesterone to inhibit estrus in pregnancy due to the presence of gestational corpus luteum on the ovary. Anestrus in pregnancy is a physiological phenomenon that ensures the normal development of the embryo. Due to aging, the function of the hypothalamic-pituitary-gonadal axis is reduced in animals, resulting in a decrease in the secretion of pituitary gonadotropins, or a decrease in the ovarian response to these hormones.

5.6.3 Pathological anestrus

Malnutrition, stress and ovarian dysfunction (such as persistent corpus luteum and corpus luteum cysts etc.) can inhibit estrus.

Project Ⅲ　Estrus Identification and Insemination

任務 1　母牛的發情鑑定與輸精
Task 1　Estrus Identification and Insemination of Cows

任務描述

人工授精是世界養牛業的通用繁殖方式，也是牛場生產管理的關鍵環節。作為一名牛場配種員，不僅要具備嫻熟的專業技術和豐富的生產經驗，還要有很強的責任心。如何鑑定母牛是否發情？如何規範完成輸精？

Task Description

Artificial insemination is widely used in dairy farms all over the world, and it is a key link in management. As a breeder, we are supposed to not only have the professional technology and abundant experience, but also have a strong sense of responsibility. How to identify whether cows are estrus or not? How to standardize the insemination process?

任務實施

一、母牛的發情鑑定

（一）外部觀察法

外部觀察法主要是通過觀察母牛的外部表現和精神狀態來判斷其發情情況。例如發情母牛表現為興奮不安，食慾減退，有強烈的求偶行為，外陰充血、腫脹、濕潤有透明黏液，產乳量下降等。

1. 準備工作

將母牛放入運動場或在牛舍內觀察，一般早晚各一次，可利用照相機或攝影機。

2. 檢查方法

主要通過觀察母牛的爬跨情況、外陰部的腫脹程度及黏液的狀態，進行綜合分析判斷。

3. 結果判定

發情母牛表現為食慾下降、興奮不安，大聲哞叫，四處走動，爬跨或接受爬跨（圖3-8）。外陰腫脹，陰道黏膜潮

Task Implementation

1　Estrus identification of cows

1.1　External observation

External observation is mainly used to judge the estrus status of cows by observing their external performance and mental state. For example, estrus cows show excitement, loss of appetite, strong courtship behaviors, pudendal congestion, swelling and moist with transparent mucus, milk production decline, etc.

1.1.1　Preparations

Observe the cows in the stadium or cowshed, usually once in the morning and once in the evening with cameras.

1.1.2　Examination method

Make a comprehensive analysis and judgment by observing the mounting behaviours, along with vulva swelling and mucus status.

1.1.3　Results judgement

Estrus cows show decreased appetite, increased excitement, loud barking, walking around, mounting behavious (Figure 3-8). The vulva is swollen, the vaginal mucosa is flushed, the uterus cervix is opened,

紅，子宮頸開張，外陰流出牽縷樣或玻璃棒狀黏液（圖 3-9）。

mucus outflow is like a stretch or glass rod (Figure 3-9).

圖 3-8　母牛發情時的爬跨行為
Figure 3-8　Mounting behavior of estrous cows

圖 3-9　發情母牛陰戶排出的黏液
Figure 3-9　Mucus excreted from the vulva of estrous cows

（二）直腸檢查法

1. 準備工作

將待檢母牛牽入固定欄內，牛尾拉向一側，使肛門充分外露。檢查人員指甲剪短磨光，帶上長臂手套，並塗抹少量的潤滑劑。

2. 檢查方法

檢查人員站在母牛的正後方，手指併攏呈錐形，旋轉且緩慢伸入肛門，進入直腸並將宿便清出。然後，將手掌向骨盆腔底部下壓找到子宮頸，手輕握子

1.2　Rectal examination

1.2.1　Preparations

The assistant leads the cow into the fixer and pulls the tail to one side, so as to let the anus fully exposed. The inspectors cut and polish nails, wear gloves with long arms and apply a small amount of lubricant.

1.2.2　Examination method

The inspector should stand right behind the cow, keep fingers together in a conical shape, rotate and extend slowly into the anus. Clean up the stool in the rectum, and then press the palm down to the bottom of the pelvic cavity to find the cervix. After finding the

宮頸向前滑動到角間溝，繼續向前找到卵巢，觸摸其大小、形狀、質地以及卵泡的發育情況（圖 3-10）。

cervix, gently hold the cervix and slide forward into the angular sulcus, continue to find out ovaries and feel the size, shape, texture, and follicle development (Figure 3-10).

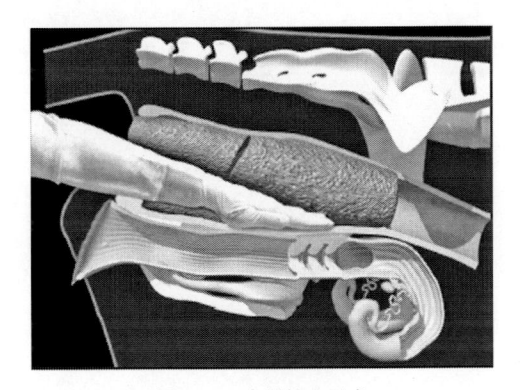

圖 3-10　直腸檢查示意
Figure 3-10　Rectal examination

3. 結果判定

母牛在間情期，一側卵巢較大，能觸到一個枕狀的黃體突出於卵巢的一端；當母牛進入發情期以後，則能觸到有一個黃豆大的卵泡存在，這個卵泡由小到大，由硬到軟，由無波動到有波動。由於卵泡發育，卵巢體積變大，直腸檢查時容易摸到。為了便於區分，確定最佳的輸精時間，通常將卵泡發育分為四個時期（圖 3-11）。

1. 2. 3　Results judgment

During diestrus, cows have larger ovaries on one side than the other, and a pillow-shaped corpus luteum could be touched protruding from one end of the ovary. During estrus, a large follicle likes a soybean could be felt, which will change from small to large, hard to soft, and without wave to with wave at the end. Ovarian volume becomes larger due to follicular development, which is easily palpable during rectal examination. In order to distinguish and determine the best time for insemination, follicular development is usually divided into four phases (Figure 3-11).

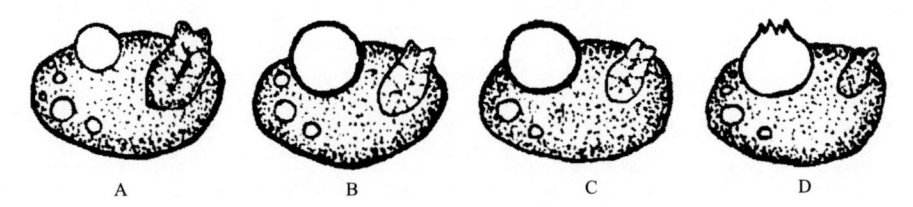

圖 3-11　牛卵泡發育模式
Figure 3-11　Model of follicular development in cow
A. 卵泡出現期　B. 卵泡發育期　C. 卵泡成熟期　D. 排卵期
A. follicular appearance　B. follicular development　C. follicular maturation　D. ovulation

第一期（卵泡出現期）：卵泡稍增大，直徑為 0.5～0.75cm，直腸觸診為一硬性隆起，波動不明顯。此期母牛已有發情表現，約持續 10h。

第二期（卵泡發育期）：卵泡直徑增大到 1～1.5cm，並突出於卵巢表面，呈小球狀，波動明顯。此期母牛處於外部表現的盛期，持續 10～12h。

第三期（卵泡成熟期）：卵泡不再繼續增大，但卵泡液增多，卵泡壁變薄，緊張度增強，直腸觸診時有「一觸即破」的感覺，似熟葡萄。

第四期（排卵期）：卵泡成熟破裂，卵泡液流出，卵巢上留下一個小的凹陷。排卵後 6～8h 可摸到肉樣感覺的黃體，其直徑為 0.5～0.8cm。

母牛最佳的輸精時間一般選擇在卵泡成熟期之後，此期母牛性慾減退，沒有明顯的發情徵狀。

（三）塗蠟筆法

塗蠟筆法鑑定發情母牛，塗抹部位為尾椎上面，從尾部到十字部，長度 30～40cm（圖 3-12）。每天塗蠟筆 1～2 次，以早晨為佳。

發情母牛接受其他母牛爬跨後，尾部毛髮被壓，上面的蠟筆塗料被摩擦掉，或者被其他母牛腹部黏附的牛糞汙染，顏色變淺、變深。未發情母牛不接受其他母牛爬跨，尾部毛髮直立或高聳，蠟筆塗料新鮮，與新塗抹的保持一致。

（四）計步器辨識法

計步器辨識法是基於發情動物的運動量要顯著高於未發情動物的原理，

Phase Ⅰ(follicular appearance): The follicle is enlarged slightly, with a diameter of 0.5-0.75 cm. The rectal palpation indicates a rigid bulge with no obvious fluctuation. The cows in this period show signs of estrus, and it lasts for about 10 hours.

Phase Ⅱ(follicular development): The diameter of follicles is 1.0-1.5 cm, protruding on the ovarian surface and showing a globular shape with obvious fluctuation. The cows in this period are at the peak of external performance, which lasts about 10-12 hours.

Phase Ⅲ(follicular maturation): The size of follicles stops growing, but follicular fluid increases. The follicular wall becomes thinner, tension increases and rectal palpation indicates a「break-upon-touching」feeling. Mature follicle shape likes a ripe grape.

Phase Ⅳ(ovulation): Matured follicular ruptures, and follicular fluid flows out, leaving a small depression on the ovary. Meat-like corpus luteum can be felt 6-8 hours after ovulation, and its diameter is about 0.5-0.8 cm.

The best insemination time for cows is usually after Phase Ⅲ, during which the cow's sexual desire decreases and has no obvious estrus symptoms.

1.3　Painting method

To identify estrus cows, the area above the cow's caudal vertebrae, from the tail to the cross, can be painted and the length should be about 30-40 cm (Figure 3-12). Repeat the painting 1-2 times a day, preferably in the morning.

If the cows are estrus, they would accept mounting and the painting area would be rubbed off or contaminated by the dung adhering to the abdomen of other cows. If the cows are diestrus, they would not accept mounting, their tail hair will be upright or towering, so the painting area stays fresh and be consistent with the new painting.

1.4　Pedometer recognition

The method of pedometer recognition is based on the principle that the movement of the estrus animal is

通過對母牛運動量的監測輔助鑑定發情母牛。計步器安裝簡單，靈敏度高，感測器識讀率高，發情監測準確，系統會根據每天的活動量自動判斷母牛的發情狀態，以此大大简化牛場繁殖人員的工作。

significantly higher than that of the non-estrus animal. The pedometer has the advantages of simple installation, high sensitivity, high sensor and high accuracy of estrus detection. The system can automatically judge the estrus state of cows according to daily activity, which greatly simplifies the work of cow breeders.

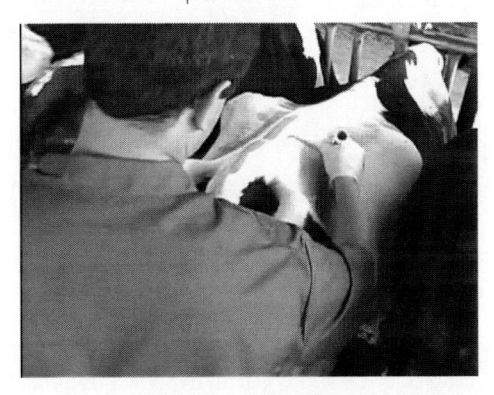

圖 3-12　塗蠟筆部位
Figure 3-12　Painting area

二、母牛的輸精

（一）準備工作

1. 母牛

母牛經發情鑑定後，將其固定，清洗、消毒外陰，尾巴拉向一側，使外陰部充分展露。

2. 器械準備

輸精器械使用前必須徹底洗淨、消毒。牛用各種輸精槍及外套管見圖3-13。外套管獨立包裝，前端圓滑，輸精時可避免劃傷子宮頸和精液倒流。

3. 精液準備

將解凍的細管精液棉塞端插入輸精槍推桿0.5cm，剪掉細管封口部，裝上鋼管套（圖3-14）。

2　Insemination of cows

2.1　Preparations

2.2.1　Cow

Pull the estrus cows into the fixer, clean and sterile its vulva, pull the tail to one side so as to fully exposed the vulva.

2.2.2　Instruments preparation

The instruments used for insemination must be thoroughly cleaned and disinfected before using. Various semen injectors and external cannulas for cows are shown in Figure 3-13. The outer cannula is packed independently and the front end of which should be smooth, thus to avoid cervical scratches and semen reflux during insemination.

2.1.3　Semen preparation

Thaw the frozen semen tubule and then insert the cotton end into the push rod of semen injector at about 0.5 cm. Cut off the sealing part of the tubule and install the steel tube cover(Figure 3-14).

4. 輸精人員準備

輸精人員穿好工作服，將指甲剪短磨平，手及手臂清洗後消毒。伸入直腸的手臂要戴一次性直腸檢查手套並塗抹潤滑劑。

2.1.4 Inseminator preparation

Inseminators should wear overalls, cut and polish their nails. Hands and arms should be cleaned and disinfected. The arm extending into the rectum should wear disposable gloves and apply lubricant.

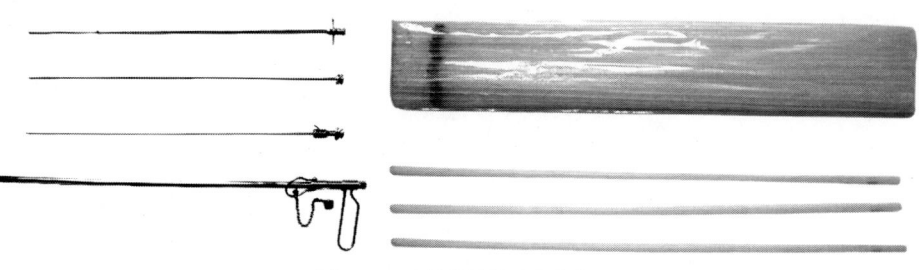

圖 3-13　牛用輸精槍及外套管

Figure 3-13　Semen injectors and external cannulas for cow

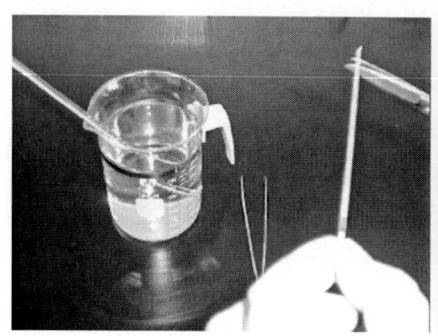

圖 3-14　精液準備

Figure 3-14　Semen preparation

（二）輸精操作

目前，牛的輸精普遍採用直腸控制子宮頸輸精法（圖 3-15）。

輸精人員站在母牛正後方，一隻手戴長臂手套，五指併攏呈錐形伸入直腸內控制住子宮頸，另一隻手持輸精器，先向斜上方伸入陰道內 5～10cm，避開尿道開口，然後兩手協同配合，把輸精器插入子宮頸內 3～4 個皺褶處，緩慢注入精液。輸精完畢後，先抽出輸精器，然後撤出手臂。

2.2 Insemination

At present, the method of rectal control of cervical insemination is widely used in cows (Figure 3-15).

The inseminator stands right behind the cow, wears long arm glove on one hand, puts five fingers together and cones into the rectum to grasp the cervix, and the other hand holds the inseminating syringe. First, insert the inseminating syringe into the vagina, which is about 5-10 cm above the oblique direction, avoiding to enter the urinary meatus. Then with two hands cooperation, insert the inseminating syringe into

專案三 發情鑑定與輸精
Project III Estrus Identification and Insemination

輸精過程中，輸精器不要握得太緊，要隨著母牛的擺動而靈活伸入；保持子宮頸呈水平狀態（圖 3-16）；避免盲目用力插入，防止損傷生殖道黏膜；輸精的原則為「輕插、適深、緩注、慢出」，防止精液倒流。

直腸控制子宮頸輸精法具有安全、母牛無痛感、受胎率高等優點，並可隔著直腸觸摸母牛子宮和卵巢的變化判斷發情和妊娠情況，防止誤配或流產。但此法初學者不容易掌握，如果操作過程中控制子宮頸的手掌位置不準，會導致輸精槍插入過淺而降低受胎率。

the cervix at 3-4 folds. While in the right position, slowly inseminate the semen. After insemination, pull out the syringe first and then the arm.

During the insemination, the inseminating syringe should not be held too tightly, it should be extended flexibly with the moves of the cow, while keeping the level position of the cervix (Figure 3-16). Don't insert the syringe blindly and forcefully to avoid genital tract mucosal damage. The principle of insemination is light insertion, moderate depth, slow injection, slow pull out and prevent reflux of semen.

Rectal control of cervical insemination has the advantages of safety operation, painless of cows, high pregnancy rate. What's more, it can be used to judge estrus and pregnancy stage by touching the uterus and ovaries of cows, so as to prevent mismating or abortion. However, this method is not easy for beginners to master. If the position of the cervix is not accurate during the operation, the insertion of the syringe would not be in the correct position which would reduce the fertility rate.

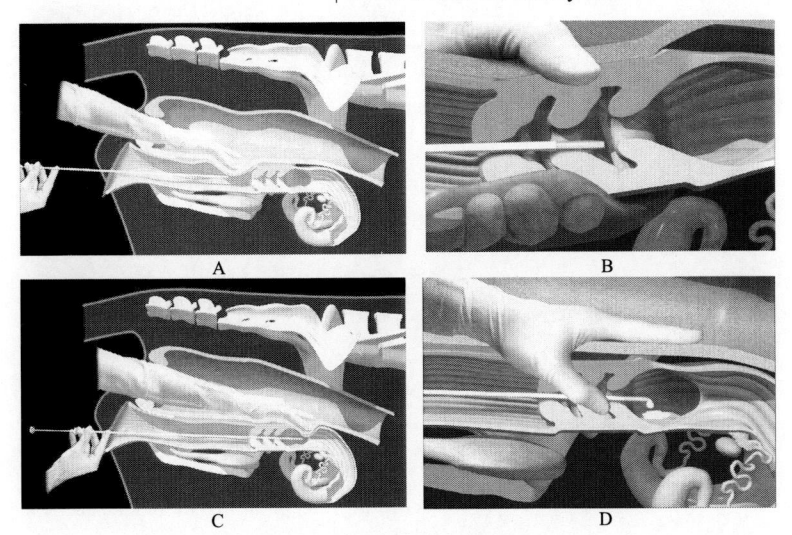

圖 3-15　直腸控制子宮頸輸精
Figure 3-15　Rectal control of cervical insemination
A. 輸精器伸入陰道　B. 輸精器進入子宮皺褶　C. 輸精器進入子宮頸口　D. 緩慢注入精液
A. insert syringe into vagina　B. insert syringe into uterine fold　C. insert syringe into cervical orifice
D. inject semen slowly

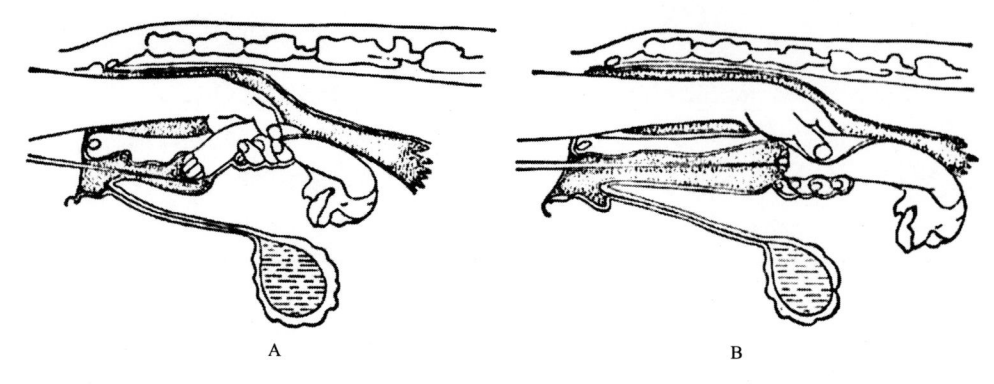

圖 3-16　控制子宮頸的操作
Figure 3-16　The operation of cervix control
A. 錯誤操作　B. 正確操作
A. incorrect operation　B. correct operation

任務 2　母羊的發情鑑定與輸精
Task 2　Estrus Identification and Insemination of Ewes

任務描述

適時而準確地把優質精液輸送到發情母羊的子宮頸口內，是保證母羊受胎、妊娠和產羔的關鍵。作為一名羊場配種員，如何進行母羊的輸精呢？

Task Description

Timely and accurate delivery of high-quality semen to the cervix of estrous ewes is the key to ensure the conception, pregnancy and lambing of ewes. As a sheep farm breeder, how can you successfully complete the insemination of ewes?

任務實施

一、母羊的發情鑑定

1. 外部觀察法

發情母羊精神興奮不安，不時地高聲「咩」叫，並接受其他羊的爬跨。同時，食慾減退、頻頻排尿，用手按壓母羊背部，母羊站立不動、擺尾，有交配慾。發情母羊的外陰部及陰道充血、腫脹、鬆弛，並有黏液流出。

Task Implementation

1　Estrus identification of ewes

1.1　External observation

The estrus ewes are agitated, crying loudly and accepting the humping of other sheep. What's more, loss of appetite, frequent urination, standing still while pressing the back with the tail wagging, with mating desire are also commonly observed. The vulva and vagina of estrus ewes are congested, swollen and relaxed, and with mucus flowing out.

2. 試情法

試情公羊的頭數為母羊數的3%～5%。在試情公羊的腹部採用標記裝置或在胸部裝上染料囊（圖3-17），每日1次或早晚各1次定時放入母羊群中，如果母羊發情並接受公羊爬跨，便將顏色印在母羊背部上。試情結束後，最好選用另一頭試情公羊，對挑出的全部發情母羊重複試情一次，準確率達90%以上。

1.2 Testing method

The rate of rams is 3%-5% of ewes. Under the abdomen of the test ram, a labeling device or a dye bag is used (Figure 3-17). Send the rams into the ewes group once a day or once in the morning and once in the evening. If the ewe is during estrus, it would accept the mounting of the ram, and the color will leave a trace on the back of the ewe. At the end of the test, it is better to repeat the test once for all selected ewes using another testing ram, and the accuracy rate is greater than 90%.

圖 3-17　羊的試情兜布
Figure 3-17　Testing cloth

3. 陰道檢查法

陰道檢查法是通過開腔器檢查母羊陰道內變化來判定母羊是否發情。該法操作簡單、準確率高，但工作效率低，適於小規模飼養戶應用。

檢查時，將待檢母羊固定，清洗外陰，擦乾，用酒精棉球消毒。將開腔器前端閉合，緩慢插入母羊陰道，輕輕打開前端，借助光線觀察陰道內的變化。發情母羊陰道黏膜潮紅、充血、表面光亮濕潤，有透明黏液滲出，子宮頸口鬆弛、開張、呈深紅色。未發情母羊陰道

1.3 Vaginal examination

Vaginal examination is to determine whether the ewe is estrus by examining the changes in the vagina of the ewe with an opener. The method is simple and high in accuracy, but it is inefficient and only suitable for small-scale feeders.

During examination, fix the ewe, clean the vulva, dry it and disinfect with alcohol. Close the fore-end of the opener and insert it into the vagina of the ewe slowly. After entering in the vagina, open the opener gently and observe the vagina by light. The vaginal mucosa of estrus ewe is flushed, congested, bright and moist, with transparent mucus exudation, and the cervical orifice is loose, open and dark red. The vaginal

黏膜蒼白、乾澀，開腔器伸入有阻力，子宮頸口關閉。檢查後稍微合攏開腔器前端，抽出。

二、母羊的輸精

母羊輸精有開腔器輸精、陰道輸精和腹腔鏡子宮內輸精等方法，目前生產上普遍採用開腔器輸精法。

助手倒提母羊，輸精員將消毒開腔器塗抹潤滑劑，旋轉插入母羊陰道內，打開開腔器，找到子宮頸外口，另一隻手持輸精器沿開腔器插入子宮頸口內0.5～1cm 緩慢注入精液（圖3-18），然後撤出輸精器，並將半開半閉狀態的開腔器慢慢取出。最後輸精員輕拍母羊的腰背部，防止精液倒流。

mucosa of the diestrus ewe is pale and dry, and the opener is obstructed and the cervical orifice is closed. After the examination, slightly close the fore-end of the opener and pull it out.

2 Insemination of ewes

There are several methods for ewe insemination, such as opener insemination, vaginal insemination and laparoscopic insemination. At present, opener insemination is widely used in ewe production.

When the assistant lifts the ewe upside down, the inseminator rubs the sterilized opener with lubricant, rotates it into the vagina, opens the opener and finds the cervix. Insert the insemination syringe along the opener into the cervix at 0.5-1 cm and slowly inject the semen (Figure 3-18). Withdraw the syringe first and then the opener. Finally, the inseminator pats the back of the ewe lightly to prevent the reflux of semen.

圖3-18 羊的開腔器輸精
Figure 3-18 Insemination of ewe with opener

任務3 母豬的發情鑑定與輸精
Task 3 Estrus Identification and Insemination of Sows

任務描述
Task Description

發情鑑定是母豬飼養管理工作中的重中之重，熟悉母豬的發情規律，掌握正確的發情鑑定方法，以便確定配種適

Estrus identification is the most important task in the breeding and management of sows. It is necessary to understand the estrus patterns and master the estrus

期，從而提高受胎率和產仔數。作為一名豬場配種員，如何準確鑑定母豬是否發情？如何確定最佳配種時間？母豬輸精時需注意哪些事項？如何防止精液倒流？

identification of sows, so as to determine the suitable time for insemination and improve the conception rate and litter size. As a pig breeder, how to accurately identify whether the sows are estrus? How to determine the best time for insemination? What should we pay attention to during insemination? And How to prevent semen reflux?

任務實施

一、母豬的發情鑑定

1. 外部觀察法

母豬開始發情時對周圍環境十分敏感，興奮不安，食慾減退，兩耳聳立，東張西望，外陰明顯充血、腫脹，陰唇黏膜隨著發情盛期的到來，變為淡紅色或血紅色，黏液量多而稀薄，性慾趨向旺盛。隨後，母豬食慾回升，表現呆滯，陰門變為淡紅、微皺、稍乾，黏液由稀轉稠，願意接受爬跨（圖 3-19），此時母豬進入發情末期，是配種的最佳時期。

Task Implementation

1 Estrus identification of sows

1.1 External observation

Sows are very sensitive to the environment when they start estrus. They are very excited and restless, showing appetite decreasing, two ears standing up and looking around. The vulva is obviously congested and swollen. The mucous inside the labia becomes reddish or hemoglobin with the estrus peak. The mucus turns large and thin, then the sexual desire tends to be vigorous. Subsequently, the appetite of sow rises, while looking sluggish. The vagina becomes light red, dry and slight wrinkled, the mucus turns from thin to thick, willing to accept humping (Figure 3-19). This is the end of estrus, which is the best time for insemination.

圖 3-19　母豬的發情表現

Figure 3-19　Oestrus behavior of sows

2. 壓背反射法

壓背反射法是利用仿生學的手段鑑定母豬發情。在沒有公豬在場的情況下，檢查人員把手按壓在母豬背部（圖 3-20）

1.2 Reflex of pressing back

Reflex of pressing back is a biomimetic method to identify estrus in sows. In the absence of boars, the inspectors press their hands on the back (Figure 3-20) or

或騎跨在母豬的背部，觀察母豬的反應。也可利用試情公豬查情，檢查人員用隔欄隔開公母豬，讓公豬與母豬頭對頭，以便母豬能看到公豬並能嗅到公豬的氣味，同時按壓母豬背部（圖 3-21）。

straddle on the back of the sow to observe the response. We can also use a boar to test the sows. The inspectors separate the sows from the boars with fences, but let the sows see and smell the boars, and then check the reflex while pressing the back of sows (Figure 3-21).

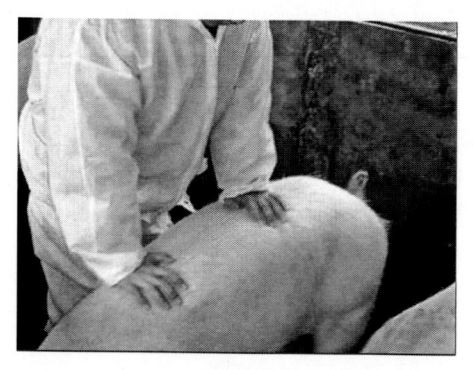

圖 3-20　豬的壓背反射
Figure 3-20　Back pressure reflex of sows

圖 3-21　公豬試情
Figure 3-21　Boar to test the sows

如果按壓背部母豬表現不安靜，前後活動，表明尚在發情初期，或者已進入發情後期，不宜配種；如果按壓後出現「靜立反射」，即母豬不哼不叫，四肢叉開，呆立不動，弓腰，這是配種最適期。如果公豬在場，發情母豬表現為立耳、翹尾，主動接近公豬，壓背時出現靜立反射。

If sows are restless and moving around after pressing, it indicates that they are still in the early stage of estrus or the late stage of estrus. It is not suitable for mating at this time. If there is a「standing reflex」after pressing, that is, the sow does not hum or cry, but divides limbs, stands still, bows waist, this is the most suitable period for mating. If boars are present, estrus sows will also show erect ears and warped tails, while approaching boars actively, and standing reflex appears when pressing back.

Project Ⅲ Estrus Identification and Insemination

母豬發情鑑定以外部觀察為主，結合壓背法進行判斷。生產實踐中，人們總結出了「一看、二聽、三算、四按背、五綜合」的母豬發情鑑定方法。即一看外陰變化、行為變化、採食情況，二聽母豬的叫聲，三算發情週期和發情持續期，四做壓背試驗，五進行綜合分析。

Estrus identification of sows is mainly based on external observation, combined with the reflex of pressing back method. In the production, the formula of「look firstly, listen secondly, count thirdly, press back fourthly and synthesize fifthly」have been summarized in the estrus identification of sows, namely: looking at vulva changes, behavior performance, feeding situation; listening to sow grunting; counting estrus cycle and estrus duration; doing back pressure test; conducting comprehensive analysis.

二、 母豬的輸精

（一）準備工作
1. 輸精器械

根據母豬情況選擇不同規格的輸精管（圖 3-22）。輸精管有螺旋頭型和海綿頭型。螺旋頭一般用無副作用的橡膠製成，適用於後備母豬的輸精；海綿頭一般用質地柔軟的海綿製成，適合於經產母豬的輸精；深部輸精利用特製的輸精導管。輸精前從密封袋中取出輸精管，手不要接觸輸精管的前 2/3 部分，在其前端塗上潤滑劑，注意不要堵塞輸精器前端的小孔。

2 Insemination of sows

2.1 Preparations
2.1.1 Equipment

We should select different sizes of insemination tubes according to sow's condition (Figure 3-22). The insemination tube has spiral head or sponge head. Spiral head is usually made of rubber without side effects and is suitable for reserve sows. Sponge head is generally made of soft sponges and is suitable for multiparous sows. Special insemination tubes is used in deep insemination. Remove the tube from the sealed bag before insemination. Pay attention not to touch the first two-thirds of the tube. Apply lubricant on the fore end of the tube, and do not block the little hole at the top.

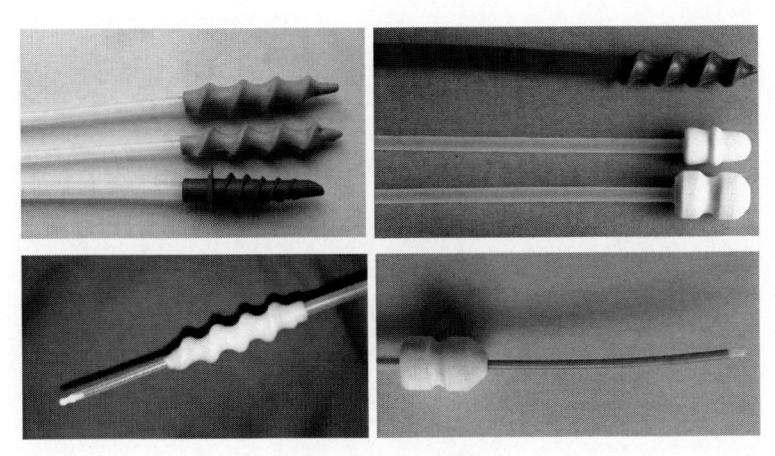

圖 3-22　豬用輸精管

Figure 3-22　Insemination tubes for sows

2. 精液

從保溫箱取出輸精瓶，輕輕顛倒幾次。輸精前，檢查精子活率，精子活率低於 0.6 的精液不能使用。

3. 母豬

用 0.1% 的高錳酸鉀溶液清洗消毒發情母豬外陰部。

(二) 輸精操作

1. 輸精時間

一般在母豬發情後 24～48h 內配種容易受胎。老齡母豬發情時間較短，排卵時間會提前，應提前配種；青年母豬發情時間長，排卵期相應後移，宜晚配；中年母豬發情時間適中，應該在發情中期配種。

生產上，一般上午發現靜立反射的母豬，下午輸精一次，第二天下午進行第二次輸精；下午發現靜立反射的母豬，第二天上午輸精一次，第三天上午再進行第二次輸精。兩次輸精時間間隔至少 8h。

2. 輸精方法

母豬的陰道部和子宮頸結合處界限不明顯，可直接將輸精管插入陰道。輸精時，將輸精膠管塗以少量潤滑劑，一隻手將母豬陰唇分開，另一隻手持輸精管呈 45°角斜上方插入母豬陰道內，避開尿道口後再平直前進。當感到向前推進有阻力時，說明海綿頭已到達子宮頸外口，然後將輸精管左右旋轉推送 3～5cm，當海綿頭插入子宮頸管內後，子宮頸管受到刺激會收縮，使海綿頭鎖定在子宮頸管內。回拉時感到有一定阻力，連接輸精瓶（袋），

2.1.2　Semen

Take out the semen from the incubator and gently turn it upside down several times. Before insemination, sperm viability should be examined. If sperm viability is below 0.6, the semen can not be used.

2.1.3　Sows

Wash and disinfect the vulva of sows with 0.1% potassium permanganate solution.

2.2　Insemination

2.2.1　Insemination time

Generally, sows are easy to conceive within 24-48 hours after estrus. Because the estrus time of old sow is short and the ovulation time will be ahead of time, so it should be mated in advance. On contrary, the estrus time of young sow is long and the ovulation period will be moved back, so it is suitable for late mating. The estrus time of middle-aged sow is moderate, so it should be mated in the middle of estrus.

In production, sows with standing reflex in the morning should be inseminated once in the afternoon, and re-inseminated in the afternoon of the next day. Those with standing reflex in the afternoon should be inseminated once in the next morning and repeat in the third day morning. The interval between two inseminations should be at least 8 hours.

2.2.2　Insemination methods

The boundary between vagina and cervix of sow is not obvious, and the semen injector can be inserted directly into vagina. During insemination, apply a small amount of lubricant to the semen injector, separate the labia of the sow with one hand, and insert the semen injector obliquely above at a 45-degree angle into the vagina with the other hand to avoid the urethral orifice, then move straight forward. When there is resistance to move forward, it means that the sponge head has reached the outer cervical orifice, and then the vas deferens are rotated and pushed forward for about 3-5 cm. When the sponge head is inserted into the cervical canal, the cervical canal will be stimu-

緩慢輸入精液（圖 3-23）。

在輸精過程中，輸精員同時按摩母豬陰戶或大腿內側，以刺激母豬的性興奮，使其子宮收縮產生負壓，促進精液吸收。輸精時不要太快，一般需 3～10min 輸完。輸精完畢緩慢抽出輸精管，並用手指按壓母豬臀部使其安靜片刻，以防精液倒流。

lated and contracted, so that the sponge head is locked in the cervical canal. When pulling back, you can feel a certain resistance, signaling the time for connecting the vase(bag) and slowly injecting semen(Figure 3-23).

In the process of insemination, the inseminator should massage the vulva or inner thigh of the sow to stimulate its sexual excitement. And this can make its uterus contract to produce negative pressure, and promote semen absorption. Insemination should not be too fast, it usually takes 3-10 minutes to complete. After insemination, vas deferens should be slowly drawn out, and press the buttocks of the sow to keep them stay for a moment, which could prevent the reflux of semen.

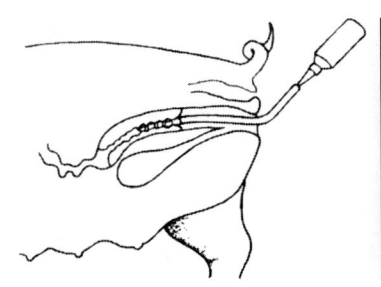

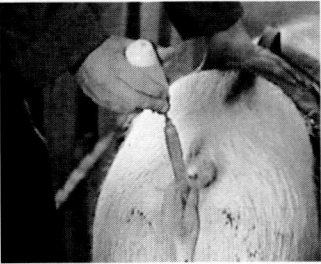

圖 3-23　母豬的輸精
Figure 3-23　Insemination of sows

任務 4　雞的人工輸精
Task 4　Insemination of Hens

任務描述

隨著中國集約化養殖的快速發展，規模化雞場普遍採用人工授精，提高了種蛋的受精率，降低了飼養成本。人工輸精是一項認真、細緻且具有技術性的工作，不同的輸精部位和深度對受精率有何影響？

Task Description

With the rapid development of intensive farm in China, artificial insemination is widely used in large-scale chicken farms, which improves the fertilization rate of eggs and reduces the feeding cost. Artificial insemination is a detailed, meticulous and highly technical work. What are the effects of different positions and depths of insemination on fertilization rate?

任務實施

一、準備工作

1. 母雞的選擇

輸精母雞應是營養中等、泄殖腔無炎症的母雞。開始輸精的最佳時間應為產蛋率達到70%的種雞群。

2. 器具及用品準備

雞用輸精器（圖2-24）、原精液或稀釋後的精液、酒精棉球等。

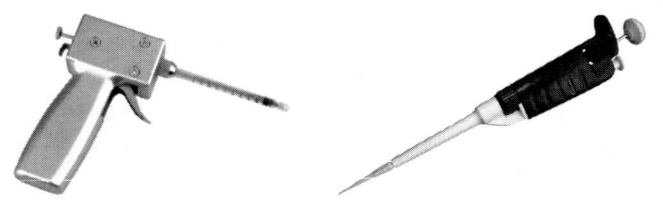

圖 3-24　雞用輸精器

二、輸精要求

雞適宜的輸精間隔為5～7d，每次輸入原精液0.025～0.03mL或稀釋精液0.1mL，輸入有效精子數至少為5 000萬個。輸精時間應選擇在大部分母雞產蛋之後進行，最好在16：00左右。

三、輸精操作

泄殖腔外翻法是輸精最常使用的方法。翻肛員右手打開籠門，左手伸入籠內抓住母雞雙腿，把雞的尾部拉出籠門口外，右手拇指與其他四指分開橫跨於肛門兩側的柔軟部分向下按壓，當給母雞腹部施加壓力時，泄殖腔便可外翻，露出輸卵管口（圖3-25）。此時，輸精員手持輸精槍對準輸卵管開口中央，

Task Implementation

1 Preparations

1.1 Hens

Fertilized hens should be medium nourished and with no inflammation in the cloaca. The best time for insemination is when the laying rate reaches 70%.

1.2 Equipment

Prepare several inseminating syringes (Figure 3-24), original semen or diluted semen, alcohol cotton ball, etc.

Figure 3-24　Inseminating syringes for hens

2 Insemination requirements

The suitable interval of insemination for hens is 5-7 days. The insemination volume should be 0.025-0.03 mL original semen or 0.1 mL diluted semen. The number of effective sperm should be at least 50 million. The insemination time should be arranged after the laying time of most hens, preferably around 16:00.

3 Insemination

Everted cloaca method is the most commonly used method for hen insemination. With right hand opening the cage, the holder reaches into the cage with his left hand, grabs the two legs of the hen and pulls the tail out of the cage door. After fixing the hen, press the soft part of lower abdomen with the right hand. It is easier if we put the right thumb separated from the other four fingers across both sides of the anus. When pressure is applied to the hen's abdomen, the cloaca can be turned outwards, exposing the oviduct (Figure 3-25). At this time, the operator could insert into the oviduct at 1-2 cm and inject the semen. When the operator is injecting the semen, the holder should release the pressure on the abdomen of the hen, thus the oviduct outlet retracted and the semen could be absorbed.

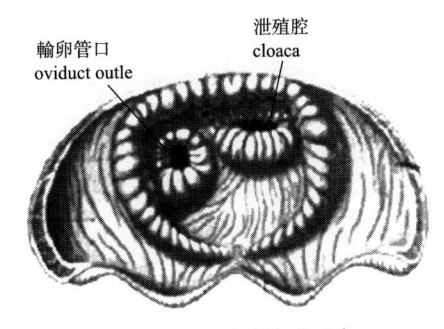

Figure 3-25　Insemination position of hens

4　Cautions

(1) Fasting and water deprivation for 2-3 hours before insemination.

(2) Grabbing hens and insemination should be done slightly, so as to minimize the fear and prevent the disturbance.

(3) If there is hard egg during insemination, the action should be slight, and syringe should be inserted into the oviduct slowly along one side.

(4) The depth should be appropriate. Generally, light laying hens use shallow vaginal insemination, i.e. inserting 1-2 cm. Medium laying hens or broiler breeders should insert 2-3 cm. When the laying rate of hens declines or the quality of semen is poor, insert 4-5 cm.

(5) To prevent cross infection, the syringe should not be reused.

(6) During laying period, the oviduct outlet can be everted easily, repeat insemination once a week can ensure a higher fertilization rate.

專案四　胚胎移植
Project Ⅳ　Embryo Transfer

✦ 專案導學

　　自然條件下，牛、馬等單胎動物通常一年產 1 胎，一生繁殖後代僅 16 隻左右；豬也不過百頭。利用胚胎移植可開發優良母畜的繁殖潛力，免去冗長的妊娠期，胚胎取出不久即可再次發情、配種和受孕，快速擴大良種畜群，因此掌握胚胎移植技術十分必要。

✦ Project Guidance

　　In nature, singleton animals such as cattle and horses usually produce one birth per year, only about 16 offsprings are born in a lifetime. And pigs have nearly 100 offsprings. Embryo transfer can develop the reproductive potential of excellent female animals, avoid the lengthy gestation period. And they could be in estrus, mating and conceived again soon after the embryo is taken out, so it is necessary to master the technology of embryo transfer.

✦ 學習目標

>>> 知識目標

• 掌握胚胎移植的概念。
• 理解胚胎移植的生理學基礎和基本原則。
• 熟悉胚胎移植的技術流程。
• 瞭解體外受精、性別控制、胚胎分割、胚胎嵌合、細胞核移植等繁殖新技術。

>>> 技能目標

• 熟練掌握母畜的同期發情和超數排卵技術。
• 規範完成牛、羊的胚胎移植操作。

✦ Learning Objectives

>>> Knowledge Objectives

• To master the concept of embryo transfer.
• To understand the physiological basis and basic principles of embryo transfer.
• To be familiar with the technological process of embryo transfer.
• To know some new reproductive technologies such as in vitro fertilization, sex control, embryo splitting, embryo chimerism, nuclear transfer, etc.

>>> Skill Objectives

• To master the technology of estrus synchronization and superovulation of female animals.
• To complete the embryo transfer of cattle and sheep standardly.

相關知識

一、胚胎移植的概念

胚胎移植俗稱「借腹懷胎」，是指將哺乳動物體內或體外生產的早期胚胎移植到同種的、生理狀況相似的雌性動物生殖道內，使之繼續發育成新個體。通常，將提供胚胎的個體稱為供體，接受胚胎的個體稱為受體。實際上，胚胎移植就是產生胚胎的供體和孕育胚胎的受體分工合作共同繁衍後代的過程，其中供體決定移植後代的遺傳特性，受體隻影響其體質發育。

二、胚胎移植的發展簡況

1890 年，英國劍橋大學的 Walter Heape 首次將純種安哥拉兔的 2 枚胚胎移植到一隻純種比利時兔的輸卵管內（用同種公兔交配後 3h），結果生出了 2 隻安哥拉仔兔和 4 隻比利時仔兔，首次證實了胚胎移植技術的可行性。

家畜的胚胎移植開始於 1930 年代。1934 年首先在綿羊上獲得成功，隨後又相繼在山羊（1949 年）、豬（1951 年）、牛（1951 年）和馬（1973 年）上取得成功。1971 年，世界上出現了第一個胚胎移植公司。1975 年 1 月，國際胚胎移植學會成立大會在美國科羅拉多州丹佛召開。目前，美國、法國、德國、澳洲、加拿大、日本等國都已建立了牛胚胎移植的商業機構。

中國家畜胚胎移植的研究始於 1970 年代，先後在兔（1973 年）、綿羊（1974 年）、牛（1977 年）、山羊（1980

Relevant Knowledge

1 The concept of embryo transfer

Embryo transfer refers to the transfer of early embryos produced in vivo or in vitro from mammals into the reproductive tract of female animals with similar physiological status, so that they can continue to develop into new individuals. Usually, the one who provides the embryo is called the donor, and the one who receives the embryo is called the receptor. In fact, embryo transfer is a cooperational process between the donor and receptor. The donor determines the genetic characteristics of the transplanted offspring, and the receptor only affects their physical development.

2 Development of embryo transfer

In 1890, Walter Heape of Cambridge University first transplanted two embryos of pure Angolan rabbit into the oviduct of a pure Belgian rabbit (3 hours after mating with the same male rabbit). As a result, two Angolan rabbits and four Belgian rabbits were born, which proved the feasibility of embryo transfer technology for the first time.

Embryo transfer of livestock began in the 1930s. In 1934, it was first succeeded in sheep, and then succeeded in goats (1949), pigs (1951), cattle (1951) and horses (1973). In 1971, the first embryo transfer company appeared in the world. In January 1975, the founding conference of the International Society for Embryo Transfer was held in Denver, Colorado, USA. At present, the United States, France, Germany, Australia, Canada, Japan and other countries have established commercial institutions for cattle embryo transfer.

The research of livestock embryo transfer in China began in 1970s. It has succeeded in rabbits (1973), sheep (1974), cattle (1977), goats (1980), and horses

年)、馬(1982年)等家畜上獲得成功，1990年代開始逐步在牛生產中推廣應用。

三、 胚胎移植的意義

1. 充分發揮優良母畜的繁殖潛力

通過胚胎移植，可使供體母畜省去冗長的妊娠期，從而縮短了繁殖週期。若再結合超數排卵技術，一次即可獲得更多的優良胚胎，大大提高了繁殖效率，尤其是對牛、羊等單胎家畜。在自然繁殖狀態下，一頭母牛平均每年只能獲得1頭犢牛，利用胚胎移植技術，一頭良種母牛一年能獲得25～30頭犢牛，最多可達50多頭。

2. 代替種畜的引進

種畜引進價格高、運輸不便、檢疫和隔離程序複雜。胚胎移植不受時間和地點的限制，這樣就可以通過胚胎的運輸代替種畜的引進，節省購置和運輸活畜的費用。此外，胚胎移植後代容易適應本地區的環境條件，也可以從養母得到一定的免疫力。

3. 保存品種資源

胚胎的冷凍保存，可以避免活畜保種因疾病、自然災害而滅絕的風險，且成本低、易實施。胚胎移植不僅保存了珍貴的遺傳資源，也為珍稀動物的繁衍和國際交流創造了機遇。

4. 克服不孕

有些優良母畜容易發生習慣性流產或難產，或者由於其他原因不宜進行妊娠過程，可以採用胚胎移植使之正常

(1982). And it has been gradually applied in cattle production since 1990s.

3 Importance of embryo transfer

3.1 Bringing into full play the reproductive potential of excellent female animals

By embryo transfer, donors can avoid long gestation period and shorten the reproductive cycle. Combined with superovulation, more excellent embryos can be obtained at one time, which greatly improves the reproductive efficiency, especially for singleton animals such as cattle and sheep. In natural, a cow can only give birth to a calf every year. By embryo transfer, a good cow can get 25-30 calves a year, up to 50 calves.

3.2 Replacing the introduction of breeding animals

The introducing of breeding animals is costly, transportation is inconvenient, procedures of quarantine and isolation are complex. Embryo transfer is not limited by time and place, so the introduction of breeding animals can be replaced by the transport of embryos, and the cost of purchasing and transporting live animals can be saved. In addition, the offsprings of embryo transfer are easy to adapt to the local environmental, and can also get certain immunity from fostress.

3.3 Preservation of variety resources

Embryo cryopreservation can avoid the risk of extinction due to diseases and natural disasters, and it is cheap and easy to implement. Embryo transfer not only preserves precious genetic resources, but also creates opportunities for the reproduction of rare animals and international exchanges.

3.4 Overcome infertility

Some excellent female animals are prone to habitual abortion or dystocia, some are not suitable to bear the pregnancy process for other reasons. By embryo transfer, they can reproduce normally.

繁殖後代。

5. 利於防疫

在養豬業中，為了培育無特定病原體（SPF）豬群，向封閉豬群引進新的個體時，為了控制疾病，往往採用胚胎移植技術代替剖腹取仔的辦法。

6. 生物學的研究手段

胚胎移植是研究受精過程、胚胎學和遺傳學等基礎理論的有效方法之一，也是胚胎分割、胚胎嵌合、體外受精、性別鑑定、核移植、基因導入等其他胚胎工程技術實施的必不可少的環節。

四、胚胎移植的生理學基礎與基本原則

（一）胚胎移植的生理學基礎

1. 母畜發情後生殖器官的孕向發育

大多數自發性排卵的動物，發情後不論是否配種，或配種後是否受精，生殖器官都會發生一系列變化，如卵巢上出現黃體、子宮內膜增生、子宮腺體發育等，為早期胚胎的發育創造良好環境。

2. 早期胚胎的游離狀態

胚胎在附植之前處於游離狀態，尚未與子宮建立組織連繫，營養主要來源於胚胎內的卵黃。這一特性是胚胎採集、保存、培養等操作的理論基礎。

3. 子宮對早期胚胎的免疫耐受性

在妊娠期內，母體局部免疫逐漸發生變化，加之胚胎表面特殊免疫保護物質的存在，受體母畜對同種胚胎和胎膜組織一般不發生免疫排斥反應。

3.5 Benefiting epidemic prevention

In pig industry, in order to breed specific pathogen free (SPF) pigs and introduce new individuals to closed pig herds, embryo transfer is often used to replace caesarean section for disease control.

3.6 Research means of biology

Embryo transfer is one of the effective methods to study the basic theories of fertilization, embryology and genetics, and it is also an essential link in the implementation of other embryo engineering technologies, such as embryo splitting, embryo chimerism, in vitro fertilization, sex identification, nuclear transfer and gene introduction etc.

4 Physiological basis and basic principles of embryo transfer

4.1 Physiological basis of embryo transfer

4.1.1 Pregnancy development of female reproductive organs after estrus

Most spontaneously ovulated animals will undergo a series of changes in their reproductive organs, whether they are mating or not after estrus, fertilized or not after mating, such as corpus luteum on the ovary, endometrial hyperplasia and uterine glands development etc., which create a good environment for the development of early embryos.

4.1.2 Free state of early embryos

The embryo is in a free state before implantation, and it has not established a tissue connection with the uterus. Its nutrition mainly comes from yolk of embryo. This characteristic is the theoretical basis of embryo collection, preservation and culture etc.

4.1.3 Immunological tolerance of uterus to early embryos

During pregnancy, due to the gradual changes of maternal local immunity and the existence of special immune protective substances on the embryo surface, the receptor generally does not have immune rejection to the same embryo and fetal membranes.

4. 胚胎遺傳物質的穩定性

胚胎的遺傳資訊在受精時就已確定，受體僅影響其體質發育，而不能改變其遺傳特性。

(二) 胚胎移植的基本原則

1. 胚胎移植前後環境的一致性

(1) 分類學上的一致性。一般來講，親緣關係較遠的物種，胚胎的生物學特性、發育條件、發育速度以及母體的子宮環境差異較大，胚胎與受體之間無法進行妊娠辨識。因此，供體和受體一般要求為同種，但這並不排除異種間（在演化史上血緣關係較近、解剖和生理特點相似）胚胎移植的可能性。

(2) 生理學上的一致性。即供體和受體在發情時間上的同期性，一般要求相差不超過 24h。

(3) 解剖部位的一致性。胚胎移植前後所處的空間環境要相似，即從供體輸卵管內採集的胚胎應移植到受體的輸卵管內，從供體子宮內採集的胚胎應該移植到受體的子宮內。

2. 胚胎的發育期限

從生理學角度講，胚胎採集和移植的期限不應超過週期黃體的壽命，最遲要在黃體退化之前數日進行。因此，胚胎採集多在配種後 3~8d 進行，受體也應在相同時間內接受胚胎移植。

3. 胚胎的品質

在胚胎移植的過程中，胚胎不應受到任何不良因素的影響，移植的胚胎必須經過鑑定確認為發育正常的胚胎。

4.1.4 Stability of embryonic genetic material

The genetic information of the embryo has been determined at the time of fertilization. The receptor only affects its physical development, but can not change its genetic characteristics.

4.2 Basic principles of embryo transfer

4.2.1 Consistency of environments before and after embryo transfer

(1) Taxonomic consistency. The biological characteristics, developmental conditions, developmental speed of embryos and maternal uterine environment differ greatly among species with distant relationships, and pregnancy can not be identified between embryos and receptors. Therefore, donors and receptors should be homologous generally, but this does not exclude the possibility of embryo transfer between different species (closely related in evolutionary history, similar in anatomical and physiological characteristics).

(2) Physiological consistency. That is to say, the estrus is synchronous between donors and receptors, generally the difference should not exceed 24 hours.

(3) Consistency of anatomical location. The spatial environment should be similar before and after embryo transfer, that is, embryos collected from the donor's oviduct should be transferred to the receptor's oviduct, and embryos collected from the donor's uterus should be transferred to the receptor's uterus.

4.2.2 Development period of embryos

In the view of physiology, the period of embryo collection and transfer should not exceed the lifespan of the cycle of corpus luteum, several days before corpus luteum regression at latest. Therefore, embryo collection is usually carried out within 3-8 days after mating, and the receptor should receive embryos at the same time.

4.2.3 Quality of embryos

In the process of embryo transfer, embryos should not be affected by any harmful factors. The transferred embryos must be identified as normally developed.

4. 經濟效益或科學價值

應用胚胎移植技術時，應考慮成本和最終收益。通常，供體胚胎應具有獨特的經濟價值，如生產性能優異或科學研究價值重大。

五、 胚胎移植的操作程序

（一）供體和受體的選擇

1. 供體的選擇

供體應該是良種母畜，具有較高的育種價值和良好的繁殖性能，對超數排卵反應處理好。經產母畜應在生殖機能恢復正常後方可作為供體。

2. 受體的選擇

受體可選用生產性能一般，但應具有良好的繁殖性能和健康狀態。

（二）供體的超數排卵

在母畜發情週期的適當時期，利用外源促性腺激素進行處理，從而增加卵巢的生理活性，誘發多個卵泡同步發育成熟並排卵。

（三）供體的配種

為了獲得較多發育正常的胚胎，對供體配種時應使用精子活率高、精子密度大的優質精液，適當增加輸精次數，輸精間隔為8～10h。

（四）胚胎的採集

胚胎的採集又稱為沖胚，是利用沖胚液將早期胚胎從子宮或輸卵管內沖出並收集的過程。胚胎採集分為手術法和非手術法兩種，前者適用於各種家畜，後者僅適用於牛、馬等大家畜，且只能

4.2.4 Economic or scientific value

When applying embryo transfer technology, the cost and benefit should be considered. In general, embryos should have unique economic value, such as excellent production performance or great scientific research value.

5 Operational procedures for embryo transfer

5.1 Selection of donors and recipients

5.1.1 Selection of donors

The donor should be an excellent female animal, with high breeding value and good reproductive performance. And it reacts well to superovulation. The multiparous dams should not be used as a donor until the reproductive function returns to normal.

5.1.2 Selection of recipients

Recipients with moderate production performance can be selected, but it should have good reproductive performance and good health.

5.2 Superovulation of donors

In the proper period of estrus cycle of female animals, exogenous gonadotrophin was used to treat them, which can increase the physiological activity of ovary, and induce multiple follicles to mature synchronously and ovulate.

5.3 Mating of donors

In order to obtain more normal embryos, excellent semen with high viability and high density should be used for donor mating. The number of inseminations should be increased appropriately and the interval between inseminations should be 8-10 hours.

5.4 Embryo collection

Embryo collection, also known as embryo flushing, is the process of using flushing fluid to rinse early embryos out of the uterus or oviduct and collect them. There are two methods for embryo collection: surgical method and non-surgical method. The former is suitable for all kinds of livestock, while the latter is

在胚胎進入子宮角以後進行。各種家畜的胚胎發育速度見表 4-1。採集胚胎的時間一般在配種後 3～8d，發育至 4～8 細胞或 8 細胞以上為宜，牛最好在配種後 6～8d，此時胚胎發育至桑葚胚或者早期囊胚，便於非手術法採集和移植。

only used for cattle and horses etc, and it can only be carried out after the embryo enters the uterine horn. The embryo development speeds of various animals are shown in Table 4-1. The time of embryo collection is usually 3-8 days after mating, and it is suitable to develop to 4-8 cells or more. The best time for cattle is 6-8 days after mating, when the embryo develops to morula or early blastocyst, and it is convenient for non-surgical collection and transplantation.

表 4-1 各種家畜的胚胎發育速度（排卵後天數）
Table 4-1 Embryo development speeds of various animals (days after ovulation)

動物 Animal	2 細胞 2 cells	4 細胞 4 cells	8 細胞 8 cells	16 細胞 16 cells	進入子宮 Entering the uterus	囊胚形成 Blastocyst formation	附植 Implantation
牛 Cattle	1～1.5	2～3	3	3～4	4～5	7～8	22
綿羊 Sheep	1.5	2	2.5	3	3～4	6～7	15
豬 Pig	1～2	2～3	3～4	4～5	5～6	6	13
兔 Rabbit	1	1～1.5	1.5～2	2～3	2.5～3	3～4	

1. 手術法

手術法多用於羊、豬和兔等小動物。即通過外科手術在腹中線做一切口，輕輕拉出子宮角和輸卵管，根據胚胎的發育階段，選擇不同的沖胚方法（圖 4-1）。

（1）輸卵管採胚法。當胚胎處於輸卵管時（排卵後 1～3d），採用此方法。對於羊、兔，用帶有磨鈍針頭的針筒刺入子宮角尖端，注入沖胚液，在輸卵管傘部接取；對於豬，則可反向沖取。

5.4.1 Surgical method

It's mostly used in small animals such as sheep, pigs and rabbits. Through surgical incision in the ventrimeson, the uterine horn and oviduct are pulled out gently. We choose different methods of rinsing embryo according to the development stage of embryo (Figure 4-1).

(1) Embryo collection from oviduct.

This method is used when the embryo is in the oviduct (1-3 days after ovulation). For sheep and rabbits, a syringe with a blunt needle is inserted into the tip of the uterine horn, injecting flushing fluid and picking up the embryo at the oviduct umbrella. For pigs, flushing can be reversed.

（2）子宮角採胚法。當確認胚胎已進入子宮角內，可採用此法，即從子宮角基部注入沖胚液，在子宮角尖端接取，或者反向沖洗。

(2) Embryo collection from uterine horn.
This method is used when the embryo has entered the uterine horn. That is, injecting flushing fluid from the base of the uterine horn and picking up the embryo at the tip of uterine horn, or back washing.

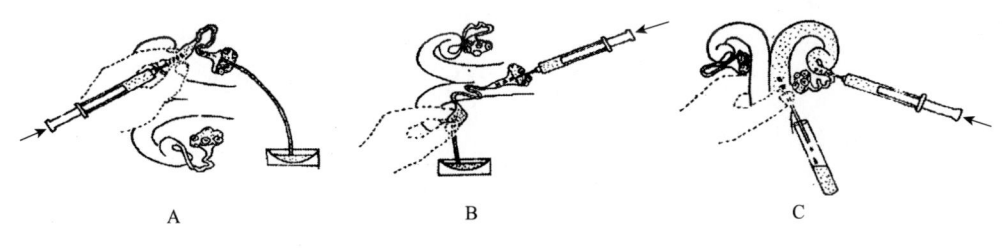

圖 4-1　手術法沖胚示意

Figure 4-1　Surgical flushing embryos

A. 由子宮角向輸卵管傘沖洗　B. 由輸卵管傘向子宮角沖洗　C. 子宮角沖洗

A. flushing from uterine horn to oviduct umbrella　B. flushing from oviduct umbrella to uterine horn

C. flushing from uterine horn

2. 非手術法

對於牛、馬等體型較大的動物，一般採用非手術法。非手術法回收胚胎都是在配種後 6～8d 進行，比手術法簡單易行，對生殖器官的傷害較小。

（五）胚胎的檢查

胚胎的檢查是指將回收的胚胎置於體視顯微鏡下檢查胚胎的數量和品質，並進行等級分類。不同發育階段的正常牛胚胎見圖 4-2。

生產中常用形態學方法進行胚胎的級別鑑定，通過觀察胚胎的形態、卵裂球的大小與均勻度、色澤、細胞密度、與透明帶間隙以及細胞變性等情況，將胚胎分為 A、B、C、D 四個等級，具體標準見表 4-2。

5.4.2 Non-surgical method

For large animals such as cattle and horses, non-surgical methods are generally used. Non-surgical embryo collection is carried out 6-8 days after mating, which is simpler and less harmful to reproductive organs than surgical method.

5.5 Embryo examination

Embryo examination is to check the quantity and quality of the recovered embryos under a stereomicroscope and grade them. The normal bovine embryos at different developmental stages are shown in Figure 4-2.

Embryos are classified into four grades: A, B, C and D by observing the morphology of embryos, the size and evenness of blastomere, color, cell density, clearance with zona pellucida and cell degeneration. The standards are shown in table 4-2.

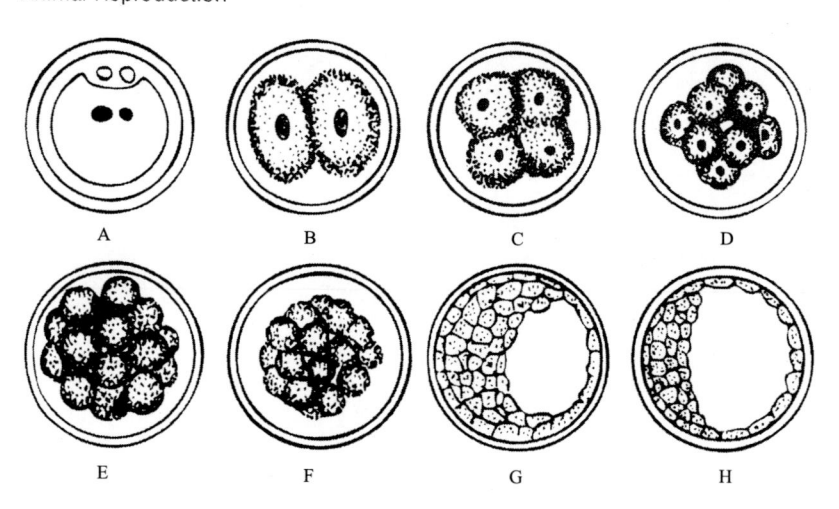

圖 4-2 不同發育階段的正常牛胚胎

Figure 4-2　Embryo development in the bovine

A. 1 細胞期（0～2d）　B. 2 細胞期（1～3d）　C. 4 細胞期（2～3d）　D. 8 細胞期（3～5d）
E. 桑葚胚（6～7d）　F. 緊實桑葚胚（6～8d）　G. 早期囊胚（6～8d）　H. 擴張囊胚（8～9d）
A. 1 cell (0-2 days)　B. 2 cells (1-3 days)　C. 4 cells (2-3 days)　D. 8 cells (3-5 days)
E. morula (6-7 days)　F. tight morula (6-8 days)
G. early blastocyst (6-8 days)　H. expanded blastocyst (8-9 days)

表 4-2　胚胎分級標準

Table 4-2　Embryo grading standards

級別 Grade	標準 Standards
A	胚胎發育階段與胚齡一致，形態完整；卵裂球輪廓清晰，大小均勻，結構緊湊，細胞密度大；色調和透明度適中，基本無游離細胞，變性細胞比例＜10% The embryo development stage is consistent with the embryonic age and the morphology is complete. The blastomere is clear in outline, uniform in size, compact in structure and dense in cell density. Tint and transparency is moderate, and it almost has no free cells, the percentage of degenerated cells is less than 10%.
B	胚胎發育階段與胚齡基本一致，形態完整；卵裂球輪廓清晰，大小基本一致，細胞結合略顯鬆散，密度較大；色調和透明度適中，變性細胞比例為 10%～20% The embryo development stage is basically aligned with the embryo age and the morphology is complete. The blastomere is clear in outline, basically same in size, slightly loose in structure and dense in cell density. Tint and transparency is moderate and the percentage of degenerated cells is 10%-20%.
C	胚胎發育階段比正常遲緩 1～2d；卵裂球輪廓不清晰，大小不均勻；色調變暗，結構鬆散，游離的細胞較多，變性細胞達 30%～40% Embryo development is 1-2 days slower than normal. The blastomere is not clear in outline and uneven in size. Tint is darkening and structure is loose, it has more free cells and the percentage of degenerated cells is up to 30%-40%.
D	胚胎發育遲緩 2d 以上，細胞團破碎，變性細胞比例超過 50% Embryo development is 2 days slower than normal, cell mass is broken and the percentage of degenerated cells is more than 50%.

（六）胚胎的保存

胚胎的保存是指將胚胎在體內或體外正常發育溫度下，暫時儲存起來而不使其活力喪失；或將其保存於低溫或超低溫條件下，使細胞新陳代謝和分裂速度減慢或停止，一旦恢復正常發育溫度，又能繼續發育。目前，哺乳動物胚胎的保存方法較多，主要包括常溫保存、低溫保存和超低溫冷凍保存等。

1. 常溫保存

將胚胎置於 15～25℃ 的培養液中保存，此溫度下胚胎能保存 24h，僅用於短暫的保存和運輸。

2. 低溫保存

將胚胎置於 0～10℃ 保存，此時，胚胎卵裂暫停，代謝減慢。不同動物的胚胎對溫度的反應不同，幾種哺乳動物的適宜保存溫度分別為：小鼠 5～10℃、家兔 10℃、綿羊 10℃、山羊 5～10℃、牛 0～6℃、豬 15～20℃ 為宜。

3. 超低溫冷凍保存

將胚胎置於超低溫環境中（－196℃）保存，其新陳代謝及發育暫時停止，可對胚胎進行長期保存。常用的方法有逐步降溫法、快速冷凍法及玻璃化冷凍法。快速冷凍法需使用專門的冷凍儀，適合大規模胚胎生產使用。玻璃化冷凍法是將胚胎放入含高濃度抗凍劑的冷凍保存液中，通過快速降溫使胚胎形成玻璃化狀態，此法操作簡便快速，但必須嚴格控制每一步操作環節。

（七）胚胎的移植

同採集胚胎一樣，胚胎的移植也有手術法和非手術法兩種，前者適於羊、

5.6 Embryo preservation

Embryo preservation is the temporary storage of embryos at normal developmental temperature in vivo or in vitro without loss of viability, or the storage of embryos at low or ultra-low temperatures, which slows down or stops the metabolism and division of cells. And the embryos can continue to develop once the normal developmental temperature is restored. At present, there are many ways to preserve mammalian embryos, including normal-temperature preservation, low-temperature preservation and cryopreservation.

5.6.1 Normal-temperature preservation

Embryos are preserved in nutrient solution at 15-25℃. At this temperature, the embryos could be preserved for 24 hours, which is only for short-term storage and transportation.

5.6.2 Low-temperature preservation

Embryos are preserved at 0-10℃. At this time, the cleavage of the embryo pauses and the metabolism slows down. Different animals have different responses to temperature variations. The suitable temperature is 5-10℃ in mice, 10℃ in rabbits, 10℃ in sheep, 5-10℃ in goats, 0-6℃ in cattle and 15-20℃ in pig.

5.6.3 Cryopreservation

Embryos can be preserved for a long time by putting them in ultra-low temperature(－196℃), and their metabolism and development are temporarily stopped. There are three methods: stepwise cooling, rapid freezing and vitrification. Rapid freezing needs special embryo freezer, which is suitable for large-scale embryo production. Vitrification is to put embryos into the cryopreservation solution with high concentration of antifreeze and make embryos vitrified by rapid cooling. This method is simple and fast, but every step of operation must be strictly controlled.

5.7 Embryo transfer

Like collecting embryos, there are two methods for embryo transfer: surgical method and non-surgical meth-

豬、兔等體型較小動物，後者僅適於牛、馬等大家畜。在胚胎移植前，應檢查受體母畜排卵一側卵巢上的黃體數量及發育情況，只有黃體發育良好的受體才能用於胚胎移植。

六、 胚胎移植存在的主要問題

1. 超數排卵效果不穩定

超數排卵並不能每次都達到預期效果，不同批次的藥品存在效價差異，不同的個體和年齡對超數排卵處理的反應差異很大，造成排卵率極不穩定。

2. 胚胎回收率較低

家畜在超數排卵處理後，排卵數過多往往會降低胚胎的回收率，其原因可能是由於卵巢在外源激素的作用下，體積增大，使輸卵管傘難以完全包被卵巢，造成一些卵子丟失。一般情況下，胚胎的回收率在50%～80%。

od. The former is suitable for small animals such as sheep, pigs and rabbits, and the latter is only suitable for large animals such as cattle and horses. Before embryo transfer, the number and development of corpus luteum on the ovary with ovulation of the recipient should be checked. Only the recipients with well-developed corpus luteum can be selected for embryo transfer.

6　Main problems in embryo transfer

6.1　The effect of superovulation is unstable

Superovulation can not achieve the desired effect every time. There are differences in the potency of drugs in different batches, and the response of different individuals and ages to superovulation varies greatly, so the ovulation rate is extremely unstable.

6.2　The recovery of embryos is low

After superovulation, excessive ovulation often reduces the recovery rate of embryos. The ovaries are enlarged with exogenous hormones, which makes it difficult for the oviduct umbrella to completely encapsulate the ovary, resulting in the loss of some ovums. Generally, the recovery rate of embryos is between 50% and 80%.

任務1　羊的胚胎移植
Task 1　Embryo Transfer in Sheep

任務描述

羊肉是集營養和保健於一體的肉食品，越來越受到消費者的青睞。目前，肉羊的供種體系不健全，種羊引進價格昂貴，在一定程度上制約了肉羊產業的快速發展。隨著胚胎移植技術的大力推廣，它在肉羊的純種繁育、快速擴群及性能提高等方面具有重要作用。如何進行羊的胚胎移植？

Task Description

Mutton is fine meat for its nutrition and health-preserving properties, and enjoying growing popularity with consumers. At present, the supply system of breeding sheep is not perfect, and the introduction price of breeding sheep is expensive, which restricts the rapid development of sheep industry to some extent. With the extension of embryo transfer, it plays an important role in sheep breeding, rapid expansion and performance improvement. How to carry out embryo transfer in sheep?

任務實施

一、準備工作

(1) 動物。供體母羊、受體母羊。

(2) 儀器。超淨工作臺、恆溫水浴鍋、體視顯微鏡、內窺鏡、CO_2培養箱、手術臺等。

(3) 材料。常規手術器械、平皿、量筒、移液器、細管(1/4)、針筒(20G)、0.5％利多卡因、孕酮陰道栓(CIDR)、沖胚液、FSH、LH、PMSG、HCG、氯前列烯醇、青黴素、鏈黴素、蒸餾水、水溫計等。

二、操作方法

(一) 供、受體羊的選擇

供體羊應健康無病，生產性能高，無遺傳缺陷疾病。受體羊應選1～3歲的雜交一代或當地母羊，胎次為1～3胎，體格接近供體羊，健康無病，無繁殖障礙。

(二) 同期發情與超數排卵

在供體羊發情週期第12天或第13天開始肌內注射FSH(150～300IU)，以日遞減劑量連續注射3d，每次間隔12h，在第5次注射FSH同時肌內注射$PGF_{2\alpha}$ 2mg。

受體羊在胚胎移植前7d注射$PGF_{2\alpha}$ 2mg，移植前6d注射FSH 50IU，每天試情並記錄發情時間。

(三) 胚胎的收集

羊的胚胎收集一般採用手術法。術前採用腹腔鏡檢查供體羊的卵巢和子宮，棄除不適宜手術的羊。

Task Implementation

1 Preparation

(1) Animals: donor and recipient ewes.

(2) Instruments: ultra-clean worktable, thermostatic water bath, stereo microscope, endoscope, CO_2 incubator, operating table.

(3) Materials: conventional surgical instruments, plate, graduated cylinder, transferpettor, tubule(1/4), injector(20G), 0.5% lidocaine, CIDR, flushing fluid, FSH, LH, PMSG, HCG, chloroprostenol, penicillin, streptomycin, distilled water, water thermometer.

2 Operation

2.1 Selection of donors and recipients

Donors should be healthy, with high production performance and no genetic defects. Recipients should be 1-3 years old hybrid or local ewes, with 1-3 calving times, close to donors in size, healthy and has no reproductive obstacles.

2.2 Estrus synchronization and superovulation

FSH(150-300 IU) is injected intramuscularly on the 12th or 13th day of the estrous cycle of the donor sheep. FSH is injected continuously for 3 days, then the dose decreases day by day, 12 hours interval each time, and $PGF_{2\alpha}$ is injected intramuscularly with a dose of 2mg at the same time of the 5th injection of FSH.

The recipient sheep are injected with 2mg $PGF_{2\alpha}$ on the 7th day and 50 IU FSH on the 6th day before embryo transfer. We test the oestrus every day and record the time of oestrus.

2.3 Embryo collection

Embryo collection in sheep usually uses surgical method. The ovaries and uterus of the donor sheep are examined by laparoscopy before operation, and the un-

1. 輸卵管採胚

採集胚胎的時間一般在配種後 2～3d，從輸卵管傘部插入細管，從宮管結合部向輸卵管方向注入 5～10mL 沖胚液。

2. 子宮角採胚

在配種後 6～7d，胚胎已進入子宮角內部。用止血鉗夾住子宮角基部，在子宮角基部插入細管，從子宮角尖端注入 50～60mL 沖胚液。

（四）胚胎的檢查

將回收液置於無菌室靜置 10～20min，吸出上清液，在體視顯微鏡下觀察胚胎的形態和發育情況，收集外形整齊、大小一致、卵裂球分裂均勻、外膜完整的胚胎備用。

（五）胚胎的移植

1. 手術法

（1）術前 24h 禁食。

（2）剃毛。

（3）固定、麻醉。

（4）消毒。

（5）手術。在一側乳房基部稍前方於腹中線旁 2～3cm 處，由後向前平行切開 5～8cm，尋找並顯露子宮、輸卵管及卵巢，觀察並記錄兩側卵巢表面的黃體大小和數量。

（6）移胚（圖 4-3）。用鈍形針頭在有黃體一側子宮角上部 1/3 處扎一小孔，將裝有胚胎的移植器沿針孔插入子宮腔內約 1cm，注入胚胎，最後縫合切口。

suitable sheep were discarded.

2.3.1 Embryo collection from oviduct

The time of embryo collection is usually 2-3 days after mating. The tubule is inserted from oviduct umbrella, and 5-10 mL flushing fluid is injected from the uterotubal junction to the oviduct.

2.3.2 Embryo collection from uterine horn

The embryo has entered the uterine horn in 6-7 days after mating. The base of uterine horn is clamped with forceps hemostatic and inserted a tubule, 50-60 mL flushing fluid is injected from the tip of uterine horn.

2.4 Embryo examination

The recovered solution is placed in sterile room for 10-20 minutes, and the supernatant is sucked out. The morphology and development of embryos are observed under the stereomicroscope. The embryos with neat shape, uniform size, even blastomere and complete membrane are collected for reserve.

2.5 Embryo transfer

2.5.1 Surgical methods

（1）Fasting 24 hours before operation.

（2）Shaving.

（3）Fixing and anesthesia.

（4）Disinfection.

（5）Operation. At the slightly-frontward base of one side of the breast, beside the ventrimeson at 2-3cm, the uterus, oviducts and ovaries are found and exposed by cutting 5-8 cm parallel from the back to the front. The size and quantity of corpus luteum on the surface of the ovaries on both sides are observed and recorded.

（6）Embryo transfer（Figure 4-3）. With a blunt needle, a small hole is pierced in the upper 1/3 of the uterine horn on the side of corpus luteum. The implanter with embryo is inserted into the uterine cavity with a depth about 1 centimeter, and then the embryo is injected into the uterus. Finally, the incision is sutured.

（7）Postoperative nursing and pregnancy diagno-

專案四　胚胎移植
Project IV　Embryo Transfer

（7）術後護理和妊娠診斷。為了防止術後引發感染，需連續注射抗生素3～7d，最好單圈飼養，做好保胎工作。

sis. In order to prevent infection after operation, antibiotics should be injected continuously for 3-7 days. It is better to raise the recipient sheep in a separate enclosure for tocolysis.

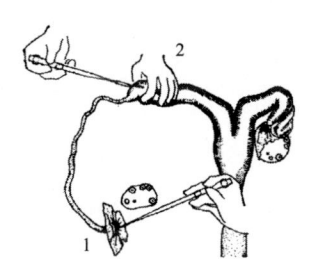

圖 4-3　羊的手術法移植胚胎
Figure 4-3　Surgical transfer of embryos in sheep
1. 將胚胎移植於輸卵管　2. 將胚胎移植於子宮角
1. embryo transfer to oviduct　2. embryo transfer to uterine horn

2. 腹腔內窺鏡移植

對羊進行全身麻醉，在乳房前方腹中線兩側各切開一個長1 cm 的小口，一側插入打孔器和腹腔鏡，另一側插入宮頸鉗。觀察卵巢上黃體的情況，選擇黃體發育良好一側的輸卵管或子宮角移植。該法對羊的損害小，整個移植過程僅需2～3 min，目前已成為羊胚胎移植的一種主要方法。

2.5.2　Endoscope transfer

Perform general anesthesia on the sheep, and cut a small opening of 1 cm in length on each side of the midline of the abdomen in front of the breast. A perforator and endoscope are inserted on one side, and cervical forcep is inserted on the other side. Observe the corpus luteum on the ovary and transfer the embryos into oviducts or uterine horn on the side with well-developed corpus luteum. This method has little damage to sheep and only takes 2-3 minutes for the whole process of embryo transfer. It has become the main method of embryo transfer in sheep.

任務2　牛的非手術法胚胎移植
Task 2　Non-surgical Embryo Transfer in Cattle

任務描述
Task Description

某農業院校舉辦畜牧獸醫人才招聘會，吸引了65家企業參加，其中4家公司招聘胚胎移植技術人員9人。這些公司主要致力於優質高產乳牛的推廣，

An Agricultural University held a job fair for animal husbandry and veterinary talent, which attracted 65 companies, and 4 of them recruited 9 embryo transfer technicians. These companies are mainly committed to

通過胚胎移植技術進行品種改良和快速擴群。目前，乳牛的胚胎移植主要採用哪種方法？如何操作？

the promotion of high-quality and high-yield cows, through embryo transfer for breed improvement and rapid expansion. At present, which method is mainly used for embryo transfer of cows? How to operate?

任務實施

一、準備工作

（1）動物。供體母牛、受體母牛。

（2）儀器。超淨工作臺、恆溫水浴鍋、乾燥箱、液氮罐、體視顯微鏡、CO_2培養箱。

（3）材料。陰道開腔器、二路式或三路式沖胚管（圖 4-4）、平皿、量筒、移液器、細管（1/4）、針筒（20 G）、靜松靈、2％利多卡因、沖胚液、犢牛血清、PMSG、FSH、LH、$PGF_{2\alpha}$、青黴素、鏈黴素、水溫計等。

Task Implementation

1 Preparation

(1) Animals: donor cows and recipient cows.

(2) Instruments: ultra-clean worktable, thermostatic water bath, drying closet, nitrogen canister, stereo microscope and CO_2 incubator.

(3) Materials: vaginal opener, two-way or three-way flushing tube (Figure 4-4), plate, graduated cylinder, transferpettor, tubule (1/4), injector (20G), xylidinothiazoline, 2% lidocaine, flushing fluid, PMSG, FSH, LH, $PGF_{2\alpha}$, penicillin, streptomycin, water thermometer, etc.

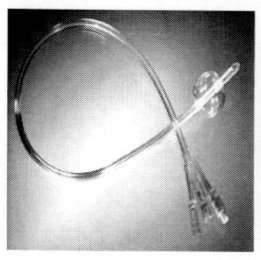

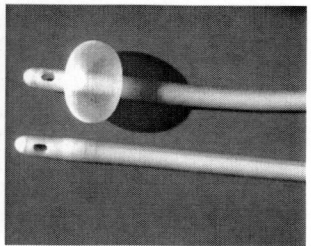

圖 4-4　沖胚管

Figure 4-4　Flushing tube

二、操作方法

（一）供、受體牛的選擇

供體牛應健康無病，種用價值高，超排效果好；受體牛可選用非良種個體，但要求體型較大、膘情中等、繁殖性能良好。

（二）供體牛的超數排卵與配種

在發情週期第 11 天肌內注射 PMSG

2 Operation

2.1 Selection of donors and recipients

Donors should be healthy, with high breeding value and good superovulation effect. Recipients can be non-breeding individuals, but with large size, moderate fat and good reproductive performance.

2.2 Superovulation and mating of donors

PMSG (3 000-4 000 IU) is injected intramuscularly on

3 000～4 000IU，或在發情週期第 11～14 天連續 4d 肌內注射 FSH，每天 2 次，間隔 12 h，總劑量 400 IU。在發情週期第 13 天肌內注射 PGF$_{2\alpha}$ 20～30 mg，促使黃體退化。供體牛在發情週期第 15 天發情，發情後 8～12 h 首次輸精，同時注射 LH 160 IU，一般輸精 3 次，間隔 8 h。

(三) 胚胎收集

母牛多採用非手術法收集胚胎（圖 4-5）。沖胚前禁水、禁食 10～24 h，將供體牛固定於固定欄內，將尾巴拉向一側，清除直腸內的宿糞，清洗陰戶周圍。

1. 麻醉

沖胚前 10 min，剪去薦椎和第一尾椎結合處或第一尾椎和第二尾椎結合處的被毛，先用酒精消毒，然後肌內注射靜松靈 3～5 mL，再用 2% 利多卡因 4 mL 進行尾椎硬膜外麻醉。

the 11th day of the estrus cycle, or FSH is injected intramuscularly between the 11th and 14th day of estrus cycle for 4 consecutive days, twice a day, at 12 hours intervals, with a total dose of 400 IU. PGF$_{2\alpha}$(20-30 mg) is injected intramuscularly on the 13th day of the estrus cycle to induce corpus luteum degeneration. The donor cattle is estrus on the 15th day of estrus cycle, and is inseminated 8-12 hours after estrus for the first time. At the same time, LH (160 IU) is injected. Insemination is performed three times at 8 hours intervals.

2.3 Embryo collection

Non-surgical method is mostly used to collect embryos in cows(Figure 4-5). Water and food are forbidden for 10-24 hours before embryo flushing. The donor cow is fixed. The tail is pulled to one side to remove the feces in the rectum and wash around the vulva.

2.3.1 Anesthesia

Ten minutes before embryo flushing, clothing hair at the junction of sacral vertebra and first coccygeal vertebra or the junction of first coccygeal vertebra and second vertebra is snipped. Disinfecting with alcohol, xylidinothiazoline is injected intramuscularly for 3-5 mL, then 4 mL 2% lidocaine is used for coccygeal vertebra epidural anesthesia.

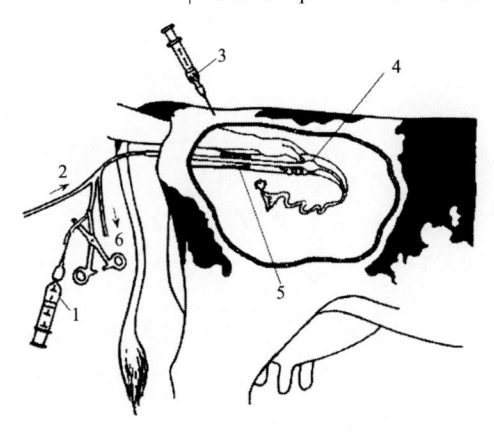

圖 4-5 牛的非手術法沖洗胚胎
Figure 4-5 Non-surgical embryo flushing in cows
1. 注氣口 2. 沖胚液入口 3. 麻醉劑 4. 氣囊 5. 子宮頸 6. 回收液出口
1. air injection port 2. import of flushing fluid 3. anesthetics 4. air sac 5. cervix 6. export of recycled liquid

2. 沖胚

組裝沖胚裝置，準備 1 000 mL 沖胚液加溫至 38℃ 備用。對於青年母牛或子宮頸通過困難的供體牛，先使用開膣器擴張子宮頸口，然後將帶有鋼芯的沖胚管插入一側子宮角大彎處，充氣 15～20 mL，氣囊脹大，使沖胚管固定於子宮基部，以免沖胚液流入子宮體並沿子宮頸口流失。抽出鋼芯，用針筒吸取沖胚液 40～60 mL 注入子宮角，回收沖胚液，反覆沖洗。用同樣方法沖洗另一側子宮角。沖洗完畢，向子宮內注入抗菌藥物。

（四）胚胎檢查

同羊的胚胎檢查。

（五）胚胎移植

1. 受體牛的篩選

受體牛在發情後 6～8 d 均可進行移植。移植前，檢查兩側卵巢的黃體發育情況，只有黃體發育合格者才能用於移植，並在臀部標記黃體所在一側。

2. 麻醉

將受體牛固定，在第 1～2 尾椎間進行硬膜外麻醉，清除宿糞，沖洗外陰部，並用酒精消毒。

3. 移胚（圖 4-6）

對照受體發情記錄，選擇適宜發育階段和級別的胚胎，封裝於 0.25 mL 塑料細管內，把裝有胚胎的細管裝入預溫的移植槍管內，金屬內芯輕輕插入細管的棉塞端內，套上滅菌的軟外套。

2.3.2 Embryo flushing

Assemble the embryo flushing device and heating 1 000 mL flushing fluid to 38℃ for use. For young cows or donors whose cervix is difficult to pass through, first use the opener to dilate the cervix. Then the flushing tube with steel core is inserted into the greater curvature of uterine horn on one side. Inflate 15-20 mL, expand the air sac, and fix the embryo flushing tube at the base of the uterus to prevent flushing embryo fluid from flowing into the uterus and losing along the cervix. Withdraw the steel core, use a syringe to draw 40-60 mL of the embryo flushing fluid into the uterine horns, recover the embryo flushing fluid, and rinse repeatedly. The other uterine horn is rinsed in the same way. After rinsing, antibio- tics are injected into the uterus.

2.4 Embryo examination

The embryo examination is the same as that of sheep.

2.5 Embryo transfer

2.5.1 Screening of recipients

The recipients can be transplanted 6-8 days after estrus. Before transplantation, check the development of the corpus luteum on both ovaries. Only those cattle with qualified corpus luteum could be used for transplantation, and the side of the qualified corpus luteum should be marked on the buttocks.

2.5.2 Anesthesia

The recipient cow is fixed and anesthetized epiduraly between the 1st and 2nd sacral vertebra. Feces are removed, vulva is washed, and disinfected with alcohol.

2.5.3 Embryo transfer (Figure 4-6)

According to the estrus records of recipients, embryos of suitable development stages and grades are selected and packed in 0.25 mL plastic tube. The tube is then put into the preheated transplantation pipe. The metal inner core is gently inserted into the cotton plug end of the tube, next the sterilized soft coat is put on.

專案四　胚胎移植
Project IV　Embryo Transfer

在胚胎移植時，用直腸控制法將移植槍通過子宮頸插入黃體側子宮角大彎處，緩慢推出胚胎，緩慢、旋轉抽出移植槍，輕輕按摩子宮角 3～4 次。

During embryo transfer, the transfer gun through the cervix is inserted into the large curve of uterine horn on the corpus luteum side by rectal grasp, and the embryo is pushed out slowly. The gun is drawn out slowly and rotationally, and the uterine horn is massaged gently for 3-4 times.

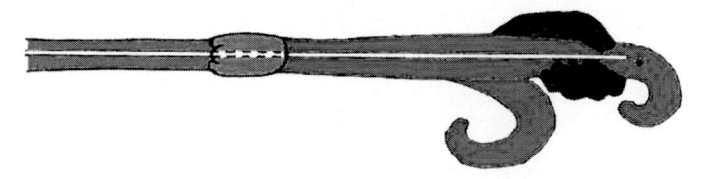

圖 4-6　牛子宮角移植胚胎

Figure 4-6　Embryo transfer from uterine horn in cows

拓展知識

一、體外受精

體外受精是動物胚胎生物工程的重要研究內容之一，是指哺乳動物的精子和卵子在體外人工控制的環境中完成受精過程的技術。早在 1878 年，德國科學家 Sckenk 就開始進行哺乳動物卵子體外受精嘗試。1951 年 Chang 和 Austin 同時發現精子獲能現象之後，哺乳動物體外受精的研究得到了飛速發展。1959 年 Chang 首次獲得「試管兔」，為哺乳動物體外受精工程奠定了基礎。到目前為止，已先後在家兔（1959 年）、小鼠（1968 年）、大鼠（1974 年）、嬰兒（1978 年）、牛（1982 年）、山羊（1985 年）、綿羊（1985 年）和豬（1986 年）等動物獲得了成功。

應用體外受精技術可獲得大量胚胎，使胚胎生產「工廠化」，為胚胎移植及相應胚胎工程提供胚胎來源，不僅在畜牧業上具有廣闊的應用前景，而且

Knowledge Expansion

1　In vitro fertilization (IVF)

In vitro fertilization is one of the important research contents of animal embryo bioengineering. It refers to the technology of mammalian sperm and ovum completing fertilization process in an artificial controlled environment in vitro. As early as 1878, Sckenk, a German scientist, attempted in vitro fertilization of ovum. After Chang and Austin discovered sperm capacitation in 1951, in vitro fertilization of mammals has developed rapidly. In 1959, Chang first obtained the「test-tube rabbit」, which laid the foundation for mammalian in vitro fertilization. So far, it is successful in rabbits (1959), mice (1968), rats (1974), infants (1978), cattle (1982), goats (1985), sheep (1985) and pigs (1986) etc.

A large number of embryos can be obtained by in vitro fertilization, which makes mass embryo production possible and provides embryos for embryo transfer and embryo engineering. It not only has broad application pros-

在醫學上可以治療不孕症，同時對於豐富受精生物學的基礎理論也具有重大意義。

（一）卵母細胞的採集

1. 離體卵母細胞採集

母畜被屠宰後，30 min 內無菌採集卵巢，用生理鹽水沖洗 2～3 次，置於 30～35℃ 的滅菌生理鹽水或 PBS 液中，盡快運回實驗室，時間以不超過 4 h 為宜。在無菌條件下，採用抽吸法、切割法或剝離法採集卵巢表面的卵母細胞。

2. 活體卵母細胞採集

藉助超音波探測儀或腹腔鏡，經陰道穿刺，直接從活體動物的卵巢中吸取卵母細胞。牛常用超音波探測儀輔助取卵，操作者將探頭插入陰道穹隆部，一隻手伸入直腸控制卵巢，緊貼在探頭所在的部位，藉助超音波圖像引導，另一隻手持吸卵針經陰道壁穿刺吸取卵母細胞。綿羊和豬等小動物常用腹腔鏡取卵。

（二）卵母細胞的體外成熟

從卵巢上採集的卵母細胞尚未成熟，需要進一步培養才能與精子受精。用於卵母細胞體外成熟的培養液有多種，其中應用最廣泛的是 TCM-199。研究發現，在培養液中加入 cAMP、生殖激素（如 FSH、17β-雌二醇、HCG 等）、血清、生長因子等，可促進卵母細胞成熟，提高培養效果。卵母細胞一般在 39℃、5% CO_2、飽和濕度條件下需培養 20～30 h。卵丘細胞充分擴張，第一極體釋放，從形態學上即可認為卵母細胞達到成熟。

pects in animal husbandry, but also can be used in clinical treatment of infertility, and has great significance for enriching the basic theory of fertilization biology.

1.1 Oocyte collection

1.1.1 In vitro oocyte collection

After slaughter, the ovaries of the female animals are collected sterilly within 30 minutes, washed with physiological saline for 2-3 times, and placed in sterilized physiological saline or PBS solution at 30-35℃. It was returned to laboratory as soon as possible for no more than 4 hours. Under sterile conditions, oocytes on ovarian surface are collected by suction, cutting or peeling.

1.1.2 In vivo oocyte collection

With the help of ultrasound detector or laparoscope, oocytes are directly extracted from the ovaries of living animals by vaginal puncture. Ultrasound detectors are often used to assist in oocyte retrieval in cattle. The operator inserts the probe into the vaginal fornix, one hand reaches into the rectum to grasp the ovary, and the other holds the oocyte suction needle through the vaginal wall to puncture and suck the oocytes. In small animals such as sheep and pigs we usually use laparoscopy.

1.2 In vitro maturation of oocytes

Oocytes collected from ovaries are immature and need further culture to fertilize with sperm. There are many kinds of nutrient solution for oocyte maturation in vitro, among which TCM-199 is the most widely used. It was found that the addition of cAMP, reproductive hormones (such as FSH, 17β-estradiol, HCG, etc.), serum and growth factors in the nutrient solution could promote oocyte maturation and improve the culture effects. Oocytes usually need to be cultured for 20-30 hours at 39℃, 5% CO_2 and saturated humidity. The cumulus cells are fully expanded and the first polar body is released. Morphologically, the oocytes can reach maturation.

（三）精子體外獲能

目前，精子獲能的處理方法主要有肝素處理法、鈣離子載體法和高滲溶液處理法。肝素是一種高度硫酸化的醣胺聚醣類化合物，與精子結合後，引起Ca^{2+}進入精子細胞內部而導致精子獲能。鈣離子載體能直接誘發Ca^{2+}進入精子細胞內部，提高細胞內的Ca^{2+}濃度，從而導致精子獲能。精子表面含有許多膜蛋白，即所謂的「去能因子」。當用高滲溶液處理精子時，可促使膜蛋白脫落而使精子獲能。

（四）體外受精

將獲能精子與成熟卵母細胞共同培養完成受精過程。體外受精的培養系統主要包括微滴法和四孔培養板法兩類。微滴法是一種應用最廣的培養系統，即在培養皿中將受精液做成20～40 μL的微滴，上覆石蠟油，每滴放入成熟卵母細胞10～20枚及獲能精子（1.0～1.5）$\times 10^6$個/mL，孵育6～24 h。如出現精子穿入卵內，頭部膨大，第二極體排出，原核形成和正常卵裂等，即可確定為受精。

（五）早期胚胎的體外培養

精子和卵子受精後，受精卵需移入胚胎培養液中繼續培養至緻密桑葚胚或早期囊胚階段。許多動物早期胚胎均存在「體外發育阻滯」現象，即胚胎發育到一定時期會受到不同程度的阻滯，牛、綿羊為8～16細胞、山羊為2細胞、豬為4細胞。研究發現，在38～39℃、

1.3 In vitro capacitation of sperm

At present, the main methods of sperm capacitation are heparin treatment, calcium ionophore and hyperosmotic solution treatment. Heparin is a highly sulfating amino-polysaccharide compound, which binds to sperm and causes Ca^{2+} to enter the sperm cells, resulting in sperm capacitation. Calcium ionophore can directly induce Ca^{2+} to enter sperm cells and increase the concentration of Ca^{2+} in sperm cells, thus leading to sperm capacitation. Membrane proteins on the sperm surface are the so-called 「decapacitation factor」. When sperm were treated with hyperosmotic solution, membrane proteins could be caducous and sperm could be capacitated.

1.4 In vitro fertilization

Fertilization was completed by co-culture of capacitated sperm and mature oocyte. The culture systems of in vitro fertilization mainly include microdrop sytem and four-hole plate system. The microdrop system is the most widely used culture system. Fertilizer fluid is made into 20-40 μL microdroplets in a culture dish and covered with liquid paraffin. Each drop is placed into 10-20 mature oocytes and capacitated sperm(1.0-1.5)$\times 10^6$/mL, then incubated for 6-24 hours. If sperm penetrates the ovum, head expands, the second polar body is discharged, pronucleus forms and cleavage occurs, fertilization can be concluded as completed.

1.5 In vitro culture of early embryos

After fertilization, the fertilized ovum needs to be transferred into the embryo nutrient solution to develop into tight morula or early blastocyst stages. Many animal embryos have 「developmental block *in vitro*」, that is, embryonic development will be stunted in a certain period, cattle and sheep are at 8-16 cells stage, goats are at 2 cells stage, pigs are at 4 cells stage. It was

5%CO_2、飽和濕度、碳酸氫鈉或TCM-199培養條件下，採用與其他細胞共同培養的方法可促進早期胚胎的體外發育。這類細胞的類型很多，如卵丘顆粒細胞、成纖維細胞、滋養層細胞、黃體細胞等，但仍存在著囊胚發育率低的問題，有待於進一步改善培養條件。

二、複製技術

在生物學中，複製是指由一個細胞或個體以無性繁殖的方式產生遺傳物質完全相同的一群細胞或一群個體。在動物繁殖學中，它是指不通過精子和卵子的受精過程而產生遺傳物質完全相同的新個體的一種胚胎生物技術。

（一）胚胎分割

胚胎分割是運用顯微技術人工製造同卵雙胎或同卵多胎的技術，是擴大良種胚胎來源的一條重要途徑，其理論依據是早期胚胎的每一個卵裂球都具有獨立發育成新個體的全能性。

不同階段的胚胎，分割方法略有差異。桑葚胚之前的胚胎，因卵裂球較大，直接分割對卵裂球損傷較大，常採用卵裂球分離方法；桑葚胚或囊胚，卵裂球結合緊密，細胞界限逐漸消失，多採用胚胎切割方法。在進行囊胚分割時，要注意將內細胞團等分。胚胎分割技術已在多種動物取得成功，但仍存在很多問題，需作深入研究。例如，初生重小，遺傳不完全一致，異常與畸形等。

found that co-culture with other cells could promote the development of early embryos in vitro under the conditions of 38-39℃, 5% CO_2, with saturated humidity, sodium bicarbonate or TCM-199. There are many types of these cells, such as cumulus granulosa cells, fibroblasts, trophoblast cells, luteal cells and so on. But there is still a problem of low blastocyst development rate which needs to be further improved.

2 Cloning technique

In biology, cloning refers to a cell or individual that produces a group of cells or individuals with the same genetic material through asexual reproduction. In animal reproduction, it refers to an embryo biotechnology that produces new individuals with identical genetic material without fertilization of sperm and ovum.

2.1 Embryo splitting

Embryo splitting is a technique to manually produce identical twins or polyembryony by microtechnique. It is an important way to expand the source of improved embryos. Its theoretical basis is that each blastomere of morula has the totipotency of developing independently into a new individual.

There are slight differences in the splitting methods for embryos in different developmental stages. Before morula, direct splitting has greater damage to the bigger blastomeres, so separation of blastomeres is often used. For morula or blastocyst, the blastomere is closely connected and the cell boundary gradually disappears, so embryo bisection is mostly used. During blastocyst spltting, inner cell mass should be equally divided. Embryo splitting is successful in many animals, but there are still many problems to futher study. For example, small birth weight, genetic inconsistency, abnormalities and deformities etc.

(二) 核移植

1996 年，英國科學家 Wilmut 等人利用綿羊的乳腺細胞成功複製了「桃莉」，這一劃時代的科技成果震動了整個世界，引起了生物學相關領域的一場革命。核移植的基本操作程序見圖 4-7。

1. 供體核的分離

胚胎複製的供體核來自早期胚胎，將供體胚胎分散成單個卵裂球。體細胞複製的供體核為體外傳代培養的體細胞，經血清飢餓使細胞處於 G_0 期。

2.2 Nuclear transplantation

In 1996, British scientist Wilmut successfully cloned Dolly from the mammary gland cells of sheep. This epoch-making scientific and technological achievement shocked the whole world and caused a revolution in biology and related fields. The basic procedures of nuclear transplation are shown in Figure 4-7.

2.2.1 Isolation of donor nucleus

The donor nuclei of embryo cloning come from the early embryos that are dispersed into single blastomeres. The donor nuclei of somatic cell cloning are somatic cells subcultured in vitro. Serum starvation keeps the cells in G_0 phase.

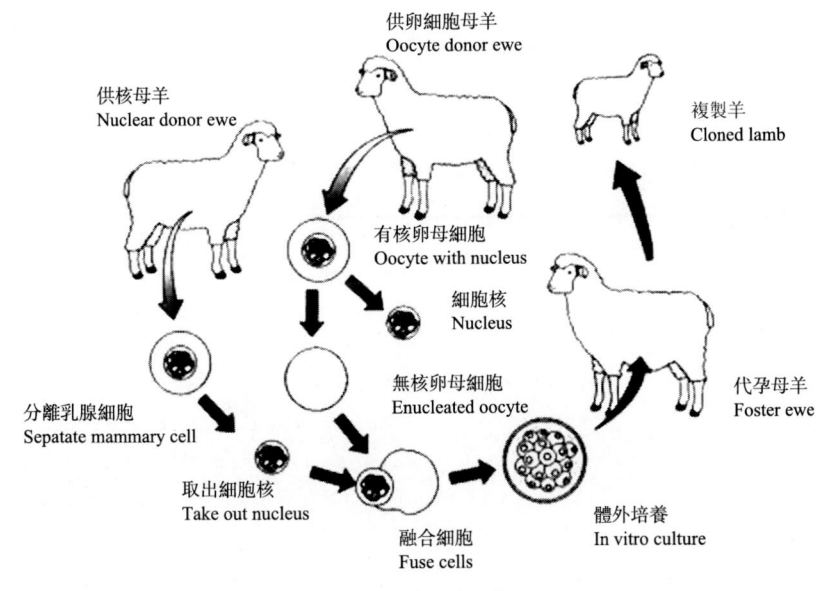

圖 4-7 核移植程序

Figure 4-7 The procedures of nuclear transplation

2. 卵母細胞的去核

將超數排卵回收的卵母細胞或體外培養成熟的卵母細胞置於含細胞鬆弛素 B 和秋水仙胺的培養液中，通過顯微操作儀，用吸管吸出第一極體以及處於細

2.2.2 Enucleation of oocyte

The oocytes recovered from superovulation or matured oocytes in vitro are placed in nutrient solution containing cytochalasin B and colchicine. The first polar body, chromosomes at metaphase of cell division

胞分裂中期的染色體和周圍的部分細胞質。

3. 細胞融合

將一個卵裂球或體細胞注入去核卵母細胞的卵黃周隙或細胞質中進行融合。細胞融合方法有電融合法和仙臺病毒法，後者因融合效果不穩定，且具有感染性，已逐漸被捨棄。電融合具有活化卵母細胞和誘導細胞融合雙重作用，即在一定強度的電脈衝作用下，供受體相鄰界面的細胞膜發生穿孔，形成細胞間橋從而達到融合，融合率可達 96%。

4. 重構胚活化

在正常受精過程中，精子穿過透明帶觸及卵黃膜時，引起卵子內鈣離子濃度升高，卵子恢復正常的細胞週期，啟動胚胎發育，這一現象稱為活化。在核移植過程中，由於受體卵母細胞處於細胞分裂中期，如果沒有活化刺激，則不能恢復細胞週期。人工活化重構胚常利用化學活化法和電活化法。

5. 重構胚的培養與移植

重構胚經體外或中間受體培養至桑葚胚或囊胚，回收胚胎，再進行冷凍保存或胚胎移植。

核移植技術的成功是胚胎工程技術的重大突破，雖然目前複製動物的成功率較低，但應用前景十分廣闊。通過核移植技術，可提高生產效率，加快育種進展，加強瀕危動物的保護力度，開展癌細胞相關基因活動的研究。

and some cytoplasm around them are sucked out by a straw under micromanipulator.

2.2.3 Cell fusion

A blastomere or somatic cell is injected into perivitelline space or cytoplasm of the enucleated oocyte for fusion. Cell fusion methods include electrofusion and Sendai virus. The latter has been abandoned gradually because of its unstable fusion effect and infectivity. Electrofusion can activate oocyte and induce cell fusion. Under the electric pulse of certain intensity, the membrane of the donor-recipient interface is perforated and forms intercellular bridges for fusion. The fusion rate is 96%.

2.2.4 Reconstituted embryo activation

During normal fertilization, when sperm passes through the zona pellucida and touches the plasma membrane, the concentration of Ca^{2+} in the ovum increases and the ovum resumes its normal cell cycle, so embryo development is initiated. This phenomenon is called activation. During nuclear transplantation, because of oocyte at metaphase in cell division, the cell cycle can not be restored without activation stimulus. Artificial activation of reconstituted embryos often includes chemical activation and electrical activation.

2.2.5 Culture and transplantation of reconstituted embryos

The reconstituted embryos are cultured to morula or blastocysts in vitro or through the middle receptor before being recovered, and then cryopreserved or transferred.

The success of nuclear transplation is a great breakthrough in embryo engineering technology. Although the success rate of animal cloning is low at present, its application prospect is very broad. Nuclear transplation can improve production efficiency, speed up breeding progress, strengthen the protection of endangered animals, and carry out research on gene activity related to cancer cells.

三、性別控制

性別控制是人為干預動物的生殖過程，使雌性動物產出人們期望性別後代的技術。性別控制可充分發揮受性別限制的性狀的生產潛力，加快育種進程，防止性連鎖疾病的發生。

（一）X、Y精子的分離

精子分離的主要依據是X、Y精子具有不同的物理性質（體積、密度、電荷、運動性）和化學性質（DNA含量、表面雄性特異性抗原）。

1. 流式細胞儀分離法

流式細胞儀法是目前比較科學、可靠、準確的精子分離方法。其理論依據是：X精子和Y精子的DNA含量不同，用螢光染料染色時，DNA含量高的精子吸收的染料就多，發出的螢光也強，反之發出的螢光就弱。具體方法為：先用DNA特異性染料對精子進行活體染色，然後使精子連同少量稀釋液逐個通過雷射束，探測器可探測精子的發光強度並把不同強弱的光信號傳遞給電腦。電腦指令液滴充電器使發光強度高的液滴帶正電，弱的帶負電。然後帶電液滴通過高壓電場，不同電荷的液滴在電場中被分離，進入兩個不同的收集管，正電荷收集管為X精子，負電荷收集管為Y精子。研究表明，家畜中X染色體的DNA含量比Y染色體高出3%～4%。

3 Sex control

Sex control is a technology that intervenes in animal reproductive process artificially so that females can produce the desired offsprings. Sex control can give full play to the potential of sex-limited character, speed up breeding process and prevent the genetically-linked diseases.

3.1 Separation of X and Y Sperm

The main basis of sperm separation is that X and Y sperm have different physical characteristics (volume, density, electric charge, mobility) and chemical characteristics (DNA content, male-specific antigen on surface).

3.1.1 Flow cytometry separation

At present, flow cytometry is a relatively scientific, reliable and accurate method for sperm separation. The theoretical basis is that the DNA contents of X and Y sperm are different. When using fluorescent dyes, sperm with high DNA content absorbs more dyes and emits stronger fluorescence, conversely, the fluorescence is weak. Specific methods are as follows: first, sperm is dyed using supravital staining method with DNA specific dyes, then sperm passes through the laser beam one by one with a small amount of diluent, the detector can detect the luminous intensity of sperm and transmit different optical signals to the computer. The computer instructs the droplet charger to make the droplets with high luminous intensity charged positively and weak negatively. Then charged droplets are separated into two different collecting tubes by high voltage electric field, X sperm is in positive charge collection tube and Y sperm is in negative charge collection tube. Studies have shown that the DNA content of X chromosome in livestock was 3%-4% higher than that of Y chromosome.

2. 免疫學分離法

免疫學分離法的原理是 Y 精子質膜攜帶 H-Y 抗原，而 X 精子無此抗原，利用 H-Y 抗體和 H-Y 抗原免疫反應檢測 Y 精子，再通過一定的程序分離 X 精子和 Y 精子。該法的分離效果不理想，目前很少用。

(二) 控制受精環境

染色體理論並非性別決定機制的全部，外部環境中的某些因素也是性別決定機制的重要條件，例如營養、pH、溫度等。由於 X、Y 兩類精子在子宮頸內的游動速度不同，因此到達受精部位與卵子結合的優先順序不同。另外，Y 精子對酸性環境的耐受力比 X 精子差，當生殖道 pH 呈弱酸性時，Y 精子活力減弱，失去較多與卵子結合的機會，故後代雌性較多。

(三) 胚胎的性別鑑定

哺乳動物早期胚胎的性別鑑定技術現已成熟，鑑定的準確率比較高。目前，早期胚胎性別鑑定最有效的方法主要有核型分析法和分子生物學法。

1. 核型分析法

該法主要操作程序：取少量胚胎細胞，用秋水仙素處理使細胞處於有絲分裂中期，經固定和染色處理，通過顯微鏡確定其性染色體類型是 XX 還是 XY。此法準確率可達 100%，但對胚胎損傷大，且耗時費力，難以在生產中應用。

2. 分子生物學法

此法的理論依據是 SRY 基因僅存在於 Y 染色體上，利用分子生物學技

3.1.2 Immunology separation

The principle is that Y sperm plasma membrane carries H-Y antigen, but X sperm does not. Y sperm is detected by immune reaction of H-Y antibody and H-Y antigen, then X sperm and Y sperm are separated by a certain procedure. The effect of this method is not ideal, and is rarely used at present.

3.2 Controlling fertilization environment

Chromosome theory is not the whole mechanism of sex determination. Some factors of external environment are also important for sex determination, such as nutrition, pH and temperature etc. Because of the different swimming speeds of X and Y sperm in the cervix, the priority order of reaching the fertilized site to combine with the ovum is different. In addition, the tolerance of Y sperm to acidic environment is worse than that of X sperm. When the reproductive tract pH is weakly acidic, the vitality of Y sperm decreases and more opportunities to combine with ovum are lost, so the offspring are mostly females.

3.3 Sex identification of embryos

The technology of sex identification of early embryos is fully developed, and the accuracy rate is high. At present, the most effective methods for sex determination of early embryos are karyotype analysis and molecular biology.

3.3.1 Karyotype analysis

Main operating procedures: take a few embryonic cells and treat with colchicine to make them in the metaphase of mitosis. After fixation and dyeing, the type of sex chromosome is determined by microscope as either XX or XY. The accuracy rate can reach 100%, but it causes great damage to embryos and is time-consuming, which makes it difficult to apply in production.

3.3.2 Method of molecular biology

The theoretical basis is that SRY gene only exists on Y chromosome, and the sex of embryo can be deter-

術鑑別胚胎細胞是否存在 SRY 基因即可判斷出胚胎性別。該法快速、靈敏、簡便、準確，對胚胎損傷較小，已廣泛應用於胚胎的性別鑑定。

四、胚胎嵌合

胚胎嵌合就是通過顯微操作，把兩枚或多枚胚胎融合成為一枚複合胚胎，由此而發育成的個體稱為嵌合體。胚胎嵌合的方法主要分為卵裂球聚合法和囊胚注入法，前者是將 2 個以上胚胎的卵裂球相互融合，形成一個胚胎；後者是把一個胚胎的內細胞團注入另一個胚胎的囊胚腔內，使之與原來的內細胞團融合在一起。

目前，已獲得鼠、兔、綿羊、山羊、豬、牛等種內嵌合體，以及大鼠-小鼠、綿羊-山羊、牛-水牛和鵪鶉-雞的種或屬間嵌合體。但嵌合體動物的表型性狀僅限於一代，不能傳遞給後代。

五、胚胎幹細胞

胚胎幹細胞（ESC）是早期胚胎或原始生殖細胞經體外分化抑制培養而獲得的可以連續傳代的發育全能性細胞系。目前，ESC 已經引起廣大學者的關注，對 ESC 的研究也取得了很大進展，相繼建立了小鼠（1981 年）、倉鼠（1988 年）、豬（1990 年）、水貂（1992 年）、牛（1992 年）、兔（1993 年）、綿羊（1994 年）等的 ESC 系或類 ESC 系。ESC 在功能上具有發育全能性及不斷增殖的能力，在生物學領域有著不可估量的應用價值。

mined by using molecular biology technology to identify the existence of SRY gene in embryonic cells. The method is rapid, efficient, simple, accurate and has less damage to embryos. It has been widely used in sex identification of embryos.

4　Embryo chimera

Embryo chimera refers to fusing two or more embryos into a compound embryo by micromanipulation, and the individual developed from it is called chimera. Embryo chimera includes two methods: blastomere aggregation method and blastocyst injection method, the former is to fuse the blastomeres of more than two embryos, the latter is to inject the inner cell mass of one embryo into the blastocele of another embryo to fuse.

At present, intraspecific chimeras (such as rat, rabbit, sheep, goat, pig and cattle) and interspecific chimeras (such as rat-mouse, sheep-goat, cattle-buffalo and quail-chicken) have been obtained. However, the phenotype of chimera is limited to one generation and cannot be transmitted to offsprings.

5　Embryonic stem cells

Embryonic stem cell(ESC) is a continuous totipotent cell line derived from early embryos or primordial germ cells cultured by differentiation inhibition in vitro. At present, ESC has attracted the attention of scholars, and great progress has also been made in the study of ESC. ESC lines or ESC-like lines of mice (1981), hamsters (1988), pigs (1990), mink (1992), cattle (1992), rabbits (1993) and sheep (1994) have been established successively. ESC has the totipotency and ability of continuous proliferation, and it has immeasurable applicational value in the field of biology.

1. ESC 培養體系

這是分離和培養 ESC 的關鍵環節，目前 ESC 培養體系主要有三種：條件培養體系、分化抑制因子培養體系和飼養層培養體系。

條件培養體系是將細胞培養一段時間後，回收培養液來培養 ESC。分化抑制因子培養體系是將分化抑制因子〔白血病抑制因子（LIF）或白血球介素-6〕按一定濃度直接添加到細胞培養液中培養 ESC。目前最常用的 ESC 培養體系是飼養層培養體系，飼養層一般由小鼠成纖維細胞經 γ 射線照射或絲裂黴素 C 處理獲得，與內細胞團（ICM）或原始生殖細胞（PGC）共同培養即可分離出 ESC。

2. ICM 及 PGC 的選擇和分離

許多早期胚胎都可作為建立 ESC 的材料，例如小鼠的桑葚胚、囊胚和擴張囊胚；豬、綿羊、山羊、兔、倉鼠的囊胚；牛的桑葚胚和囊胚。分離 ESC 首先要獲得 ICM，然後把 ICM 分散成單個細胞，再放入分化抑制因子培養體系中繼續培養。常用的 PGC 分離方法有機械法和消化法，對於分離的 PGC 還要進一步純化。

3. ESC 的鑑定

當 ICM 增殖後或 PGC 出現胚胎幹細胞樣複製後即可傳代，2～7 代時進行鑑定。ESC 的鑑定方法主要有形態學鑑定、表面抗原檢測、核型分析、體外分化實驗等。ESC 必須具有高度的分化潛能，被注入囊胚腔後，能參與內胚層、中胚層和外胚層的形成。體外培養時，分化誘導劑可誘導其定向分化。

5.1 ESC culture system

This is the key to isolate and culture ESC. At present, there are three main ESC culture systems: conditional culture system, differentiation inhibitor culture system and feeder layer culture system.

Conditional culture system refers to the culture of ESC by using recovery culture solution after culturing cells for a period of time. Differentiation inhibitor culture system refers to the culture of ESC by using cell culture solution which is added a certain concentration of leukemia inhibitory factor(LIF) or interleukin-6(IL-6). At present, feeder layer culture system is commonly used. Feeder layer is usually obtained from mouse fibroblasts irradiated by γ-rays or treated with mitomycin C, and then co-cultured with inner cell mass (ICM) or primordial germ cells(PGC), from which ESC can be isolated.

5.2 Selection and separation of ICM and PGC

Many early embryos can be used as materials for the establishment of ESC, such as mouse morula, blastocyst and expanded blastocyst, the blastocyst of pig, sheep, goat, rabbit and hamster, bovine morula and blastocyst. To isolate ESC, ICM should be obtained first, then ICM is dispersed into single cells and cultured in the system of differentiation inhibitors. The common separation methods of PGC are mechanical method and digestion method, the separated PGC needs to be further purified.

5.3 Identification of ESC

When ICM increases or PGC appears embryonic stem cell-like clones, it can be passaged and identified at 2-7 generations. The main identification methods of ESC include morphological identification, surface antigen detection, karyotype analysis and differentiation test in vitro. ESC must have a high differentiation potential and can participate in the formation of endoderm, mesoderm and ectoderm after being injected into blastocoele. When cultured in vitro, differentiation inducers

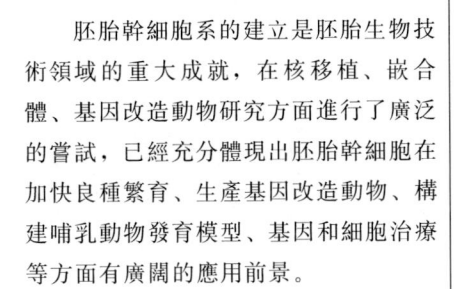

胚胎幹細胞系的建立是胚胎生物技術領域的重大成就，在核移植、嵌合體、基因改造動物研究方面進行了廣泛的嘗試，已經充分體現出胚胎幹細胞在加快良種繁育、生產基因改造動物、構建哺乳動物發育模型、基因和細胞治療等方面有廣闊的應用前景。

六、基因改造技術

基因改造技術是用一定方法將目的基因導入受體基因組中或把受體基因組中一段 DNA 切除，從而改變該物種的遺傳資訊。自 1982 年獲得基因改造「超級鼠」以來，動物基因改造技術已成為當今生命科學中一個發展最快、最熱門的領域，相繼在兔、羊、豬、牛、雞等動物獲得成功。

基因改造技術的主要步驟包括：獲取目的基因、構建表達載體、基因導入、鑑定與篩選。其中基因導入是關鍵環節，基因導入的方法主要包括顯微注射法、病毒轉染法、精子載體法、ESC 介導法及體細胞核移植法等。

1. 顯微注射法

藉助顯微操作儀，將外源基因直接注入受精卵原核。世界上第一隻基因改造小鼠就是用這種方法獲得的，該技術已廣泛應用於製作基因改造動物。

can induce its directional differentiation.

Establishment of embryonic stem cell lines is an important achievement in the field of embryonic biotechnology. Extensive attempts have been made in nuclear transplation, chimerism and transgenic animal research. It has broadened application prospects in accelerating breeding of favorable breeds, producing transgenic animals, constructing mammalian development models, gene research and cell therapy etc.

6 Transgenic technology

Transgenic technology refers to changing the genetic information of the species by either introducing the target gene into the receptor genome or removing a segment of DNA from the receptor genome. Since the transgenic super mouse was obtained in 1982, animal transgenic technology has become one of the fastest growing and highly-sought-after fields in life science. It has been successful in rabbits, sheep, pigs, cattle and chickens consecutively.

The main steps of transgenic technology include obtained the target genes, constructing expression vectors, gene introduction, identification and screening. Among them, gene introduction is the key. Its methods include microinjection, virus transfection, sperm vector, ESC mediation and somatic cell nuclear transfer.

6.1 Microinjection

With the help of micromanipulator, exogenous gene is directly injected into the pronucleus of fertilized eggs. The first transgenic mice in the world was obtained in this way. It has been widely used in the production of transgenic animals.

2. 病毒轉染法

將目的基因整合到病毒基因組中，然後利用此病毒感染胚胎細胞，即可對胚胎細胞進行遺傳轉化。該法操作簡單、宿主廣泛、轉染率高，但載體病毒基因有潛在致病性，威脅受體動物的健康安全。

3. 精子載體法

將外源基因片段與獲能精子一起孵育，通過受精過程把外源基因導入受精卵（圖4-8）。該法最大的優點是方法簡單，不需要昂貴複雜的設備，缺點是效果不穩定。

6.2 Virus transfection

Integrate the target gene into the viral genome, and use the virus to infect embryonic cells, then the embryonic cells can be genetically transformed. The method is simple in operation, wide in host and high in transfection rate, but the vector virus gene has potential pathogenicity, which threatens the health and safety of recipients.

6.3 Sperm vector

The exogenous gene fragments are incubated with capacitated sperm, and the exogenous genes are introduced into the fertilized eggs through the fertilization (Figure 4-8). It is simple and does not require expensive and complicated equipments. But it is unstable.

DNA與精子共培育
Co-incubation of DNA and sperm

受精卵細胞
Fertilized egg cell

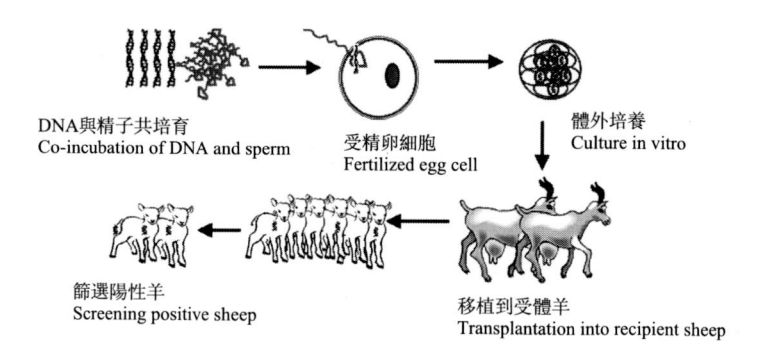

體外培養
Culture in vitro

篩選陽性羊
Screening positive sheep

移植到受體羊
Transplantation into recipient sheep

圖 4-8　精子載體法
Figure 4-8　Sperm vector

4. ESC 介導法

將外源基因整合到 ESC 基因組中的特定基因位點，通過篩選，把陽性細胞注入受體的囊胚腔內，生產嵌合體動物。當 ESC 分化為生殖幹細胞時，外源基因可通過生殖細胞遺傳給後代，在第二代獲得基因改造動物。其理論依據在於囊胚的內細胞團含有尚未分化的胚胎幹細胞，將這些幹細胞植入正常發育

6.4 ESC mediation

The exogenous gene is integrated into specific gene locus in ESC genome, and the screened positive cells are injected into the blastocele of the recipient to produce chimaera. When ESC differentiated into germ stem cells, the exogenous gene can be transmitted to offsprings, and the transgenic animals can be obtained in the second generation. Its theoretical basis is that the inner cell mass of blastocyst contains undifferentiated embryonic stem cells, implanting them into the normal

的囊胚腔之後，很快便與受體內細胞團聚集在一起，參與正常胚胎的發育。

5. 體細胞核移植法

將外源基因以 DNA 轉染方式導入能進行傳代培養的體細胞，再以這些基因改造體細胞為核供體，進行動物複製，獲得基因改造複製動物。

基因改造工程已廣泛應用於農業和醫藥，甚至與環境保護有密切的關係。在動植物生產方面，可用於進行品種改良和抗病力提高，目前已有百餘種基因改造植物進入商品化生產，例如水稻、玉米、棉花、大豆等。在醫學上，可用於基因治療和藥用蛋白生產。目前市場上的胰島素、干擾素、白血球介素-2、生長激素等基因工程藥物，均為重組蛋白質或肽類。基因工程藥物療效好、副作用小，已成為各國政府和企業投資研發的熱點。將外源正常基因導入靶細胞，以糾正或補償因基因缺陷和異常引起的疾病，從而達到治療的目的。

blastocele, they soon gather together with the recipient cells and participate in the development of normal embryos.

6.5 Somatic cell nuclear transfer

The exogenous gene is transfected into subcultured somatic cells by DNA transfection, and then these transgenic somatic cells are used as nuclear donors for animal cloning to obtain transgenic cloned animals.

Transgenic engineering has been widely used in agriculture and medicine industries, and is even closely related to environmental protection. In animal and plant production, it can be used for breed improvement and disease resistance. At present, more than 100 kinds of transgenic plants have entered commercial production, such as rice, corn, cotton, soybean and so on. In medicine, it can be used in gene therapy and pharmaceutical protein production. At present, drugs made by gene engineering such as insulin, interferon, interleukin-2 and growth hormone are all recombinant proteins or peptides. Drugs made by gene engineering have good curative effects and little side-effect, which has greatly attracted governments and enterprises to invest in research and development of this area. In order to treat diseases caused by gene defects and abnormalities, normal exogenous gene is introduced into target cells to correct or compensate for the defected genes.

專案五　妊娠診斷
Project V　Pregnancy Diagnosis

專案導學

妊娠是指母畜從受精到分娩的生理過程。母畜配種後應儘早確定其是否妊娠，通過妊娠診斷，加強對已妊娠母畜的飼養管理，使產犢間隔最小化。因此，簡便有效的妊娠診斷方法，尤其是早期妊娠診斷方法，受到廣大繁殖工作者的關注。

Project Guidance

Pregnancy is the physiological process from fertilization to delivery. Female animals should be determined whether they are pregnant or not as soon as possible after mating. Through pregnancy diagnosis, we should strengthen the feeding management of pregnant to minimize the calving interval. Thus, simple and effective methods of pregnancy diagnosis, especially early pregnancy diagnosis, are favored by breeders.

學習目標

>>> 知識目標

- 理解受精的過程。
- 掌握早期胚胎發育的特點。
- 掌握妊娠辨識和胚胎附植的過程。
- 熟悉胎膜和胎盤的結構及特點。
- 瞭解各種家畜的妊娠期生理。
- 掌握各種母畜的妊娠期。

>>> 技能目標

- 能熟練掌握各種家畜的妊娠診斷方法。
- 根據配種記錄，熟練推算各種母畜的預產期。

Learning Objectives

>>> Knowledge Objectives

- To understand the process of fertilization.
- To master the characteristics of embryonic development in the early stages.
- To master the process of pregnancy recognition and embryo implantation.
- To know the structure and characteristics of fetal membranes and placentas.
- To understand the physiology of various livestock during pregnancy.
- To master the gestation period for various livestock.

>>> Skill Objectives

- To master the diagnosis methods of pregnancy in various livestock.
- According to the mating records, estimate the due dates for various livestock.

相關知識

一、受精生理

受精是指精子和卵子結合形成合子的過程。在這一過程中，精子和卵子經歷一系列嚴格有序的形態和生理變化，單倍體的雌、雄生殖細胞共同構成雙倍體的合子。合子是新個體發育的始發點。

（一）配子的運行

在自然情況下，大多數動物的受精發生在母畜輸卵管壺腹部。精子由射精部位（或輸精部位）、卵母細胞由排出部位到達受精部位的過程，稱為配子的運行。瞭解配子在母畜生殖道內運行及其保持受精能力的時間，對確定適當的配種時間和提高受胎率具有重要的意義。

1. 精子的運行

在自然交配中，雄性動物將精液射入陰道（牛、羊、兔和靈長類動物）或子宮（馬、豬、犬和嚙齒動物）。射精後，精子在母畜生殖道的運行主要通過子宮頸、子宮和輸卵管三個主要部位，最後到達受精部位。

（1）精子在子宮頸內的運行。陰道內射精的動物，如牛、羊，在自然交配時精子存放在陰道內，部分精子藉自身運動和黏液向前運動進入子宮，大部分精子進入子宮頸隱窩的黏膜皺褶內，暫時儲存，形成精子庫，精子會隨子宮頸的收縮活動緩慢釋放進入子宮，而死精子可能因纖毛上皮的逆蠕動被推向陰道排出，或被白血球吞噬而清除。

Relevant Knowledge

1 Fertilization physiology

Fertilization refers to the process in which sperm and ovum combine to form zygotes. During this process, sperm and ovum undergo a series of strictly ordered morphological and physiological changes. The haploid germ cells constitute a diploid zygote. Zygote is the beginning of new ontogeny.

1.1 Transport of gametes

In nature, fertilization of most animals occur in the ampulla of the oviduct of the female. The process of sperm from the ejaculation site (or the insemination site) and oocyte from ovary to the fertilization site is called the transport of gametes. It is important to know the time of gametes running in the reproductive tract of female animals and their ability to maintain fertility, which is imperative for knowing proper mating time and improving conception rate.

1.1.1 Transport of sperm

In natural mating, the male ejaculates semen into the vagina (cattle, sheep, rabbit and primates) or uterus (horse, pig, dog and rodents). After ejaculation, sperm travels through the cervix, uterus and oviduct to reach the fertilization site.

(1) Transport of sperm in the cervix

Animals with endovaginal ejaculation, such as cattle and sheep, store sperm in the vagina when mating naturally. Some sperm enter the uterus by self-movement and mucus forward movement. Most sperm enter the mucosal folds of the cervical recess and are temporarily stored to form a sperm bank. Sperm release slowly into the uterus with the contraction of the cervix. Dead sperm may be pushed to the vagina through the reverse peristalsis of ciliary epithelium and expelled or cleaned up by white blood cells.

子宮頸是精子運行過程中的第一道柵欄，精子經過子宮頸篩選，既保證了運動和受精能力強的精子進入子宮，也防止過多的精子同時湧入子宮。綿羊一次射精將近 30 億個精子，但能通過子宮頸進入子宮的不足 100 萬個。

（2）精子在子宮內的運行。穿過子宮頸進入子宮的精子，在子宮肌收縮作用下，大部分進入子宮內膜腺隱窩中，形成第二個精子庫。活力較強的精子從中不斷釋放，並在子宮肌和輸卵管繫膜的收縮、子宮液的流動以及精子自身運動等綜合作用下，通過宮管連接部進入輸卵管。其中一些死精子和活動力差的精子被白血球吞噬，使精子又一次得到篩選。由於宮管連接部比較狹窄，大量精子滯留於該部，宮管連接部成為精子運行的第二道柵欄。

（3）精子在輸卵管中的運行。進入輸卵管的精子，在輸卵管的收縮、管壁上皮纖毛的擺動作用下，繼續前行。在壺峽連接部，精子因峽部括約肌的有力收縮被暫時阻擋，因此，壺峽連接部成為精子到達受精部位的第三道柵欄。各種動物能夠到達輸卵管壺腹部的精子一般不超過 1 000 個。

精子在母畜生殖道運行中的損耗情況見圖 5-1。

精子在母畜生殖道內運行的動力包括射精的力量、子宮頸的吸入作用、母畜生殖道的收縮、生殖道管腔液體的流動以及精子自身的運動等。精子由射精部位到達受精部位所需的時間與母畜生理狀況有關，一般為 20min 左右。

The cervix is the first barrier in the process of sperm transport. The cervix ensures motile and fertilized sperm enter the uterus, and also prevents excessive sperm from entering the uterus at the same time. For example a ram ejaculates nearly 3 billion sperm at a time, but less than 1 million can enter the uterus through the cervix.

(2) Transport of sperm in uterus

The sperms passing through the cervix enter the uterus. Under the contraction of uterine muscles, most of sperms enter the endometrial glands, forming a second sperm bank. Vigorous sperm is released continuously from the endometrial glands, and enters the oviduct through the uterotubal junction under the influence of the contraction of uterine muscle and mesosalpinx, flow of uterine fluid and sperm motility. The dead and poor motility sperms are swallowed by white blood cells, which further selects the viable sperm. Owing to the narrowness of the uterotubal junction, a large number of sperms remain in the uterotubal junction, which becomes the second barrier for sperm transport.

(3) Transport of sperm in the oviduct

The sperms in the oviduct move forward continuously by the contraction of the oviduct and swing of tube wall epithelial cilia. They are temporarily blocked due to the strong contraction of the isthmic sphincter in the junction of ampulla and isthmus, which becomes the third barrier for sperms to reach the fertilization site. Generally, no more than 1 000 sperms can reach the ampulla of oviduct.

The loss of sperm in reproductive tracts of female animals is shown in Figure 5-1.

The motive force of sperm in the reproductive tract of female animals includes the force of ejaculation, the absorption of cervix, the contraction of reproductive tract, the flow of liquid in the reproductive tract and the movement of sperm. The time required for sperms the fertilization site from the ejaculation site is related to the physiological conditions

精子在母畜生殖道的存活時間一般為1～2d，牛為15～56h，豬約為50h，羊約為48h，馬約為6d。精子維持受精能力的時間短於存活時間，牛約為28h，豬約為24h，綿羊為30～36h，馬為5～6d。

of the female animals, generally about 20 minutes. The survival time of sperm in the reproductive tract of the female animals is generally 1-2 days, the cow is 15-56 hours, the pig is about 50 hours, the sheep is about 48 hours, and the horse is about 6 days. The time of sperm fertilization window is even shorter than the survival time, which is about 28 hours for cattle, 24 hours for pigs, 30-36 hours for sheep, and 5-6 days for horses.

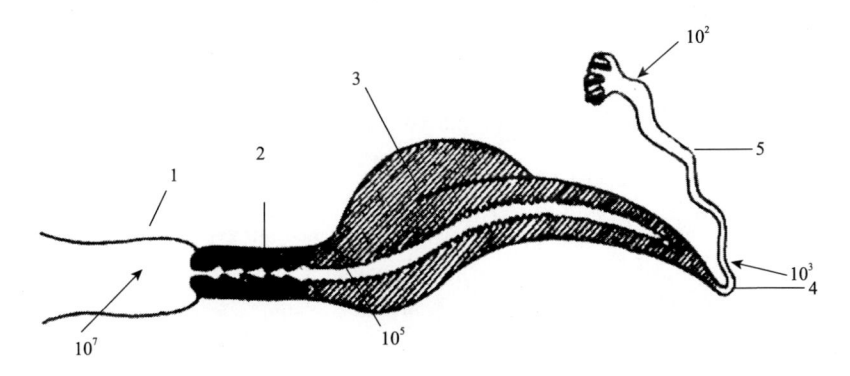

圖 5-1　精子運行中的損耗

Figure 5-1　Sperm loss in transport

1. 陰道　2. 子宮頸　3. 子宮　4. 宮管連接部　5. 輸卵管

1. vagina　2. cervix　3. uterus　4. uterotubal junction　5. oviduct

2. 卵子的運行

接近排卵時，輸卵管傘充分開放、充血，緊貼於卵巢表面。輸卵管傘黏膜上擺動的纖毛將卵巢排出的卵子接納入喇叭口。豬和馬的傘部發達，卵子易被接受；牛、羊因傘部不能完全包圍卵巢，有時造成排出的卵子落入腹腔，再靠纖毛擺動形成的液流將卵子吸入輸卵管。被傘部接納的卵子，藉輸卵管壁纖毛的顫動、平滑肌的收縮以及腔內液體的作用，被運送到受精部位。卵子在輸卵管內運行的時間，牛約為80h，豬約為50h，綿羊約為72h。但卵子保持受精能力的時間，一般為1d以內，牛為18～20h，豬為8～12h，綿羊為12～16h。

1.1.2　Transport of oocyte

Before ovulation, the oviduct umbrella is fully open and congested, and close to the ovarian surface. The waving cilia on the mucosa of the oviduct umbrella receives the oocyte from the ovary. Owing to the developed oviduct umbrellas, the oocyte of pigs and horses are easily accepted. Because of the oviduct umbrellas of cattle and sheep can not completely encircle the ovaries, sometimes the oocyte falls into the abdominal cavity, but the fluid flowing through the cilia can suck the oocyte back into the oviduct. The time of oocyte running in the oviduct is about 80 hours for cattle, 50 hours for pigs and 72 hours for sheep. However, the time window for fertilization is generally less than one day, which is 18-20 hours for cattle, 8-12 hours for pigs, and 12-16 hours for sheep.

（二）配子在受精前的準備

受精前，哺乳動物的精子和卵子都要經歷一個進一步生理成熟的階段，才能順利完成受精過程，並為受精卵的發育奠定基礎。

1. 精子的獲能

剛排出的精子沒有受精能力，只有在雌性生殖道內經歷一段時間，完成形態和生理上的一些變化，才具有受精能力，這種現象稱為精子獲能。獲能後的精子耗氧量增加，呈現一種超活化運動狀態。一般認為，精子獲能的主要意義在於使精子做好頂體反應的準備和精子超活化，促使精子穿越透明帶。

對於子宮射精型的動物，精子獲能開始於子宮，結束於輸卵管；對於陰道射精型的動物，精子獲能開始於陰道，當子宮頸開放時，流入陰道的子宮液可使精子獲能，但獲能最有效的部位是子宮和輸卵管。精子獲能也可在體外人工培養液中完成。精子在雌性生殖道內獲能的時間，因動物種類不同而異，一般牛為3～4h，豬為3～6h，綿羊為1.5h，兔為5～6h。

精子獲能是一個可逆過程。獲能的精子，若重置於精清中便又失去受精能力，這種現象稱為去能。如果去能的精子再回到母畜生殖道內，可再次獲能。在哺乳動物，去能因子無種間特異性。

獲能後的精子，在受精部位與卵子

1.2 Preparation of gametes before fertilization

Before fertilization, the sperm and oocyte of mammalians have to undergo a stage of further physiological maturation to successfully complete the fertilization process and lay a good foundation for the development of zygote.

1.2.1 Sperm capacitation

The freshly ejaculated sperm has no fertilization ability, only after a period of time in the female reproductive tract, completing some changes in morphology and physiology, it can have fertilization ability. This phenomenon is called sperm capacitation. Oxygen consumption of sperm increased after capacitation, showing a hyperactivation state. It is generally believed that the main significance of sperm capacitation lies in the preparation of acrosome reaction and sperm superactivation, and promotes sperm to pass through the zona pellucida.

In uterine ejaculation animals, sperm capacitation begins in the uterus and ends in the oviduct. In vaginal ejaculation animals, sperm capacitation begins in the vagina. When the cervix is open, the uterine fluid flowing into the vagina can make sperm capacitated, but the most effective sites of sperm capacitation are the uterus and oviduct. Sperm capacitation may also be accomplished in an artificial medium in vitro. The time of sperm capacitation in the female reproductive tract varies among species. For the cattle it is 3-4 hours, the pig is 3-6 hours, the sheep is 1.5 hours, and the rabbit is 5-6 hours.

Sperm capacitation is a reversible process. Capacitive sperm will lose fertility if they are replaced in the seminal fluid. This phenomenon is called decapacitation. If the decapacitive sperm returns to the reproductive tract of the female animal, it can be capacitated again. In mammals, decapacitation factors have no interspecific specificity.

When the capacitated sperm meets the oocyte at the

相遇，會出現頂體帽膨大，精子質膜和頂體外膜融合而形成許多泡狀結構，最後由頂體內膜和頂體基質釋放出頂體酶系，這一過程稱為頂體反應。頂體反應為精子穿越卵子並與之融合奠定了基礎。

2. 卵子的成熟

剛排出的卵子，在進入輸卵管壺腹部前尚不具備受精能力，如豬和羊排出的卵子為剛完成第一次減數分裂的次級卵母細胞，馬和犬排出的卵子僅為初級卵母細胞。卵子需在輸卵管內進一步成熟，達到第二次減數分裂中期，才具備被精子穿透的能力。同時，卵子的皮質顆粒不斷增加，並向卵的周圍移動，此時透明帶出現精子受體，卵黃膜發生亞顯微結構變化。

（三）受精過程

受精過程指精子和卵子相結合的生理過程。哺乳動物的受精過程主要包括以下 5 個階段（圖 5-2）。

fertilization site, acrosome cap enlarges, spermatozoal plasma membrane and outer acrosome membrane fuse to form many vesculation. At last, acrosome enzymes are released from inner acrosome membrane and acrosome matrix. This process is called acrosome reaction. This reaction lays the foundation for sperm to cross and fuse with the oocyte.

1.2.2 Oocyte maturation

The newly expelled oocytes do not have the ability to fertilize before entering the ampulla of the oviduct, for example, the oocytes from pigs and sheep are secondary oocytes that have just completed their first meiotic division, and the oocytes from horses and dogs are only primary oocytes. Oocytes need to mature in the oviduct to reach second meiotic metaphase before they can be penetrated by sperm. At the same time, the cortical granules of the oocytes increase continuously and move around the oocytes. At this time, sperm receptor in zona pellucida appears, and microstructural changes occur in yolk membrane.

1.3 Fertilization process

Fertilization refers to the physiological process of sperm fusing with oocyte. The fertilization process of mammals mainly includes the following five stages (Figure 5-2).

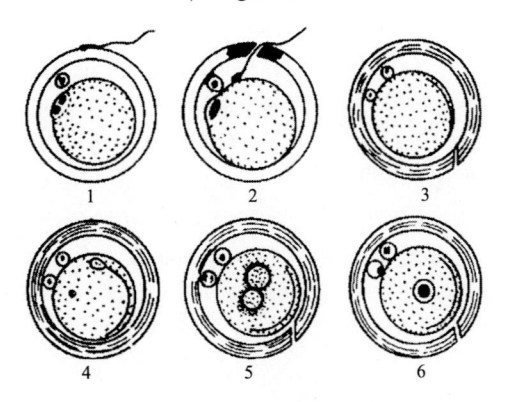

圖 5-2 受精過程

Figure 5-2　Fertilization process

1. 精子與透明帶接觸　2. 精子進入透明帶　3. 精子進入卵黃　4、5. 雄、雌原核形成　6. 配子配合

1. sperm contacts with zona pellucida　2. sperm enters zona pellucida　3. sperm enters yolk
4、5. prokaryotic formation of male and female　6. cyngamy

1. 精子穿過放射冠

放射冠是包圍在卵子透明帶外面的卵丘細胞群。受精前大量精子包圍著卵細胞，經頂體反應的精子釋放出透明質酸酶，溶解放射冠的膠樣基質，使精子接觸到透明帶。此過程需要大量精子的共同作用，此時卵子對精子無選擇性，即不存在種間特異性。

2. 精子穿過透明帶

接觸到透明帶的精子，很快與透明帶上的精子受體結合。精子受體具有明顯的種屬特異性，只有同種動物的精子才可與其受體結合。精子釋放頂體素（酶）將透明帶溶出一條通道，精子藉自身運動穿過透明帶。

當精子穿過透明帶觸及卵黃膜時，將處於休眠狀態的卵子「活化」，卵黃膜發生收縮，由卵黃釋放出某種物質，傳播到卵的表面及卵黃周隙，阻止後來的精子再進入透明帶，這一變化稱為透明帶反應。兔的卵子無透明帶反應，可在卵黃周隙內發現許多精子，這些多餘的精子稱為補充精子。

3. 精子進入卵黃

穿過透明帶的精子與卵黃膜接觸，卵黃膜表面的微絨毛包圍住精子，卵黃膜隨之與精子質膜融合，將精子「拖入卵內」。當精子進入卵黃膜時，卵黃膜立即發生卵黃緊縮、卵黃膜增厚、排出部分液體進入卵黃周隙等變化，拒絕其他精子再進入卵黃，這種現象稱為卵黃膜封閉作用或多精子入卵阻滯。

1.3.1 Sperm penetrates the corona radiata

The corona radiata is a cumulus cell group surrounding the zona pellucida. Before fertilization, a large number of sperm surround the oocyte. After the acrosome reaction, the sperm releases hyaluronidase, dissolves the colloidal matrix of the corona radiata, and makes the sperm touch the zona pellucida. This process requires the interaction of a large number of sperms, at this moment the oocyte has no sperm selectivity, that is, there is no interspecific specificity.

1.3.2 Sperm passes through zona pellucida

Sperm exposed to zona pellucida quickly binds to sperm receptors in zona pellucida. Sperm receptors have distinct species specificity and only the sperm of the same species can bind to their receptors. The sperm dissolves a channel through zona pellucida by release of acrosin(enzyme), and the sperm moves through the zona pellucida by itself.

When sperm passes through the zona pellucida and touches the yolk membrane, the dormant oocyte is activated and the yolk membrane contracts, releasing a substance from the yolk, which spreads to the surface and perivitelline space, and prevents the subsequent sperm from entering the zona pellucida. This change is called zona pellucida reaction. The oocytes of rabbits have no zona pellucida response, so many sperms can be found in the perivitelline space. These extra sperms are called complementary sperm.

1.3.3 Sperm enters the yolk

The sperm passing through zona pellucida contacts with the yolk membrane, and the microvilli on the surface of the yolk membrane embrace the sperm. The yolk membrane fuses with the sperm plasma membrane, and the sperm is dragged into the ovum. When the sperm enters the yolk membrane, the yolk membrane immediately undergoes changes, such as yolk contraction, yolk membrane getting thicker, and part of the fluid being discharged into the perivitelline space, and other spems can't enter the yolk. This phenomenon is called vitelline block.

專案五　妊娠診斷

Project V　Pregnancy Diagnosis

4. 原核形成

精子進入卵黃後，頭部膨大，尾部脫落，出現核膜和核仁，形成雄原核。精子進入卵黃後不久，卵子排出第二極體，完成第二次成熟分裂，逐漸出現核膜和核仁，形成雌原核。原核形成後，不斷髮育、變大。豬的兩個原核大小相似，其他家畜雄原核略大於雌原核。

5. 配子配合

雄原核和雌原核經充分發育，相向移動，彼此接觸，融合在一起，核仁、核膜消失，兩組染色體合併成一組，這一過程稱為配子配合。配子配合完成後，受精至此結束。

二、妊娠生理

妊娠是指母畜從受精開始，經過胚胎的生長發育，直至胎兒成熟產出體外的生理變化過程。

（一）胚胎的早期發育

從受精卵第一次卵裂至發育成原腸胚的過程，稱為胚胎的早期發育。根據形態特徵，可將早期胚胎的發育分為桑葚胚、囊胚和原腸胚三個階段（圖5-3）。

1. 桑葚胚

受精卵形成後立即在透明帶內進行卵裂。第一次卵裂，合子一分為二，形成兩個卵裂球。之後，胚胎繼續卵裂，但每個卵裂球並不一定同時進行分裂，故可能出現單數個細胞的時期。當卵裂球達到16～32個細胞時，由於透明帶

1.3.4　Pronucleus formation

When the sperm enters the yolk, the head enlarges, the tail falls off, and the nuclear membrane and nucleolus appear, forming the male pronucleus. Shortly after the sperm enters the yolk, the oocyte discharges the second polar body and completes the second mature division, and the nuclear membrane and nucleolus appears gradually, forming the female pronucleus. After the formation of pronucleus, they grow and become bigger. The male pronucleus are always slightly larger than the female pronucleus, but the two pronucleus of pigs are similar in size.

1.3.5　Cyngamy

The male and female pronucleus are fully developed, moving towards each other and fusing together, then the nucleolar and the membrane disappear. The two groups of chromosomes merge into a group. This process is called cyngamy. After cyngamy is completed, the fertilization ends.

2　Pregnancy physiology

Pregnancy refers to the physiological changes of the female animals from fertilization to the growth and development of the embryo, until the fetus matures enough to be delivered.

2.1　Early embryonic development

The process of the development from the first cleavage of a oosperm to the gastrula is called the early embryonic development. According to morphological characteristics, the early embryonic development can be divided into three stages: morula, blastocyst and gastrula(Figure 5-3).

2.1.1　Morula

The cleavage occurs immediately in the zona pellucida after the formation of the oosperm. In the first cleavage, the zygote is divided into two parts, forming two blastomeres. After that, the cleave of the embryo continues, but each blastomere does not divide at the same time, so there may be an odd number of cells ap-

的限制，卵裂球在透明帶內形成緻密的細胞團，形似桑葚，稱為桑葚胚。這一時期的變化主要在輸卵管內完成，依賴自身卵黃獲取營養。

早期胚胎的每個卵裂球都具有發育成為一個新個體的全能性，利用此特性可進行胚胎分割和移植。

peared. When the blastomeres reaches 16-32 cells, due to the restriction of zona pellucida, it forms a dense cell mass which shapes like a mulberry in the zona pellucida, hence its name morula. This period is mainly completed in the oviduct, depending on their own yolk for nutrition.

Each blastomere of an early embryo has the totipotency of developing into a new individual, which can be used for embryo splitting and transplantation.

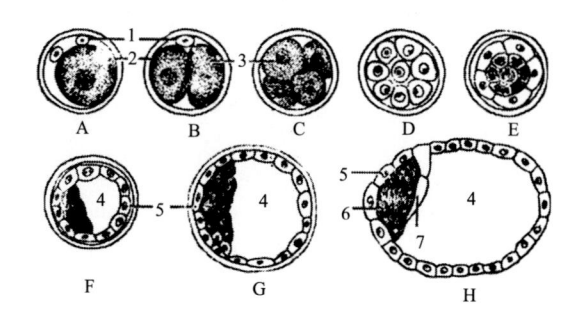

圖 5-3　受精卵的發育
Figure 5-3　Development of oosperm
A. 合子　B. 2 細胞期　C. 4 細胞期　D. 8 細胞期　E. 桑葚期　F～H. 囊胚期
1. 極體　2. 透明帶　3. 卵裂球　4. 囊胚腔　5. 滋養層　6. 內細胞團　7. 內胚層
A. zygote　B. 2 cells　C. 4 cells　D. 8 cells　E. morula　F-H. blastocyst
1. polar body　2. zona pellucida　3. blastomere　4. blastocoel
5. trophoblast　6. inner cell mass　7. endoderm

2. 囊胚

桑葚胚繼續發育，逐漸在細胞團中出現充滿液體的小腔，稱為囊胚腔，此時的胚胎稱為囊胚。隨著細胞的分裂，囊胚腔不斷擴大，最終一些細胞被擠在腔的一端，細胞密集成團，稱為內細胞團；而另一些細胞沿著透明帶的內壁排列擴展，構成囊胚腔的壁，這一單層細胞稱為滋養層。在囊胚階段，細胞開始分化，內細胞團將來發育為胎兒，滋養層將發育為胎膜和胎盤。囊胚後期，透

2.1.2　Blastocyst

The morula continues to develop, and a small cavity filled with liquid gradually appears in the cell mass, which is called a blastocyst cavity. The embryo at this time is called a blastocyst. With cell division, blastocyst cavity expands continuously, and finally some cells are squeezed into one end of the cavity, which is called inner cell mass, while others are arranged along the inner wall of zona pellucida to form the wall of blastocyst cavity, which is called trophoblast. During the blastocyst stage, the cells begin to differentiate, the inner cell mass will develop into the fetus, and the trophoblast

明帶崩解，囊胚體積迅速增大，變為擴張囊胚。囊胚階段主要從子宮乳獲取營養物質。

3. 原腸胚

囊胚進一步發育，出現了內、外兩個胚層，此時的胚胎稱為原腸胚。原腸胚繼續發育，在滋養層（即外胚層）和內胚層之間出現了中胚層，中胚層又分化為體壁中胚層和臟壁中胚層，兩個中胚層之間的腔隙，構成以後的體腔。三個胚層的建立和形成，為胎膜和胎體各類器官的分化奠定了基礎。

（二）妊娠辨識

在妊娠早期，胚胎即能產生某種化學因子（激素）作為妊娠信號傳給母體，母體隨即產生相應的生理反應，以辨識或確認胚胎的存在。由此胚胎和母體之間建立起密切的連繫，這一過程稱為妊娠辨識。妊娠辨識的實質是胚胎產生抗溶黃體物質，作用於母體的子宮或黃體，阻止或抵消 $PGF_{2\alpha}$ 的溶黃體作用，使黃體變為妊娠黃體，維持母畜妊娠。

妊娠辨識後，母畜即進入妊娠的生理狀態，但各種動物妊娠辨識的時間不同，豬為受精後 10～12 d、牛為 16～17 d、綿羊為 12～13 d、馬為 14～16 d。

（三）胚胎的附植

早期胚胎在子宮內游離一段時間後，體積越來越大，其活動逐漸受到限制，位置逐漸固定下來，胚胎的滋養層和子宮內層膜逐漸建立起組織和生理上

will develop into the fetal membrane and placenta. At the later stage of blastocyst, the zona pellucida disintegrates, and the blastocyst increases rapidly and becomes expanded blastocyst. At this stage, blastocysts obtain nutrients mainly from the uterine milk.

2.1.3 Gastrula

With the development of blastocyst, endoderm and ectoderm appear. At this time, the embryo is called gastrula. The gastrula continues to develop, mesoderm appears between trophoblast (ectoderm) and endoderm. The mesoderm differentiates into somatic mesoderm and splancchnic mesoderm. The gap between the two mesoderms forms the coelom. The formation of three parts lays a foundation for the differentiation of various organs.

2.2 Pregnancy recognition

In the early stage of pregnancy, the embryo can produce some chemical factors (hormones) as a pregnancy signal to the mother, then maternal physiological reactions occur to identify or confirm the existence of embryo. Thus, a close relationship is established between the embryo and the mother. This process is called pregnancy recognition. The essence of pregnancy recognition is that the embryo produces antiluteolytic substance, which acts on the uterus or corpus luteum and prevents or counteracts the luteolytic effect of $PGF_{2\alpha}$, transforming the corpus luteum into the corpus luteum of pregnacy, and maintains the maternal pregnancy.

After pregnancy recognition, the female enters the physiological state of pregnancy. However, the time of pregnancy recognition is different in various animals. It is 10-12 days after fertilization for the pigs, 6-17 days for cattle, 12-13 days for sheep, and 14-16 days for horses.

2.3 Embryo implantation

After a period of intrauterine dissociation, the early embryo becomes larger and larger, its activity is limited and the position is gradually fixed. The tissual and physiological links are established gradually between the trophoblast and endometrium. This process

的連繫，這一過程稱為附植。

1. 附植時間

胚胎附植是一個逐漸發生的過程，各種家畜胚胎的準確附植時間差異較大。胚胎附植的時間大體為：牛為受精後 40～45 d，馬為 90～105 d，豬為 25～26 d，綿羊為 28～35 d。

2. 附植部位

胚胎在子宮內附植時，通常都是尋找對胚胎發育最有利的位置。所謂有利，一是指子宮血管稠密的地方，可以提供豐富的營養；二是距離均等，避免擁擠。胚胎在子宮中附植的位置，因動物種類不同而異。牛、羊懷單胎時，胚胎多在排卵側子宮角下 1/3 處附植，雙胎時則均勻分布於兩側子宮角；馬懷單胎時，胚胎常遷至對側子宮角基部附植；豬的多個胚胎平均等距離分配在兩側子宮角。

（四）胎膜和胎盤

1. 胎膜

胎膜是胎兒的附屬膜，是卵黃囊、羊膜、絨毛膜、尿膜和臍帶的總稱（圖 5-4）。其作用是與母體子宮黏膜交換養分、氣體及代謝產物，對胎兒的發育極為重要。在胎兒出生後，胎膜即被摒棄，所以是一個暫時性器官。

（1）卵黃囊。哺乳動物胚胎發育初期都有卵黃囊的發育，其上分布有稠密的血管，是早期胚胎主要的營養器官，起著原始胎盤的作用。隨著胎盤的形成，卵黃囊逐漸萎縮，最後只在臍帶中留下一點遺蹟。

2.3.1 Implantation time

Embryo implantation is a gradual process, and the accurate implantation time of various livestock embryos has enormous differences. The time of embryo implantation is roughly 40-45 days after fertilization for cattle, 90-105 days for horses, 25-26 days for pigs, and 28-35 days for sheep.

2.3.2 Implantation site

When the embroy implanting in uterus, it usually looks for the most favorable position for embryonic development. The favorable position which can provide abundant nutrition is where blood vessels are dense in uterine and with equal distance from each other to avoid crowding. The implantation site of embryos in the uterus varies according to animal species. When cows and sheep have a single fetus, the embryo is often attached to 1/3 down the uterine horn on the ovulation side, twins are evenly distributed on both sides of the uterine horn. When the horses have a single fetus, the embryo is often transferred to the base of the contralateral uterine horn. When the pigs have multiples, the embryos are evenly distributed on both sides of the uterine horn.

2.4 Embryonic membrane and placenta

2.4.1 Embryonic membrane

The embryonic membrane is the accessory membrane of the fetus. It is formed of the yolk sac, amnion, chorion, allantois and umbilical cord (Figure 5-4). Its role is to exchange nutrients, gases and metabolites with maternal uterine mucosa, which is very important for fetal development. After the birth of the fetus, it is discarded, so it is a temporary organ.

(1) Yolk sac

The development of yolk sac also starts at the early age of embryo development, it is covered with dense blood vessels. It is the main nutritional organ of early embryos and plays the role of primitive placenta. With the formation of the placenta, the yolk sac atrophies gradually, leaving only a trace in the umbilical cord.

Project V　Pregnancy Diagnosis

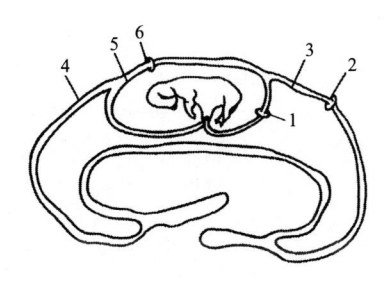

圖 5-4　豬的胎膜
Figure 5-4　Embryonic membrane of pigs
1. 尿膜羊膜　2. 尿膜絨毛膜　3. 尿膜　4. 絨毛膜　5. 羊膜　6. 羊膜絨毛膜
1. urinary amnion　2. urinary chorion　3. allantois　4. chorion　5. amnion　6. amniochorion

（2）羊膜。羊膜是包裹在胎兒外面的一層透明薄膜，在胎兒臍孔處和胎兒皮膚相連。羊膜閉合為羊膜腔，其內含有羊水，胎兒即漂浮在羊水中，對胎兒起緩衝作用。

（3）尿膜。尿膜閉合為尿囊，尿囊通過臍帶中臍尿管與胎兒膀胱相連，內含尿水，相當於胚體外的臨時膀胱。尿膜上分布有大量來自臍動脈、臍靜脈的血管。豬、牛、羊的尿囊在胎兒腹側和兩側半包圍著羊膜囊，馬的尿囊完全包裹著羊膜囊。

（4）絨毛膜。絨毛膜是胚胎的最外一層膜，表面覆蓋絨毛，嵌入子宮黏膜腺窩，形成胎兒胎盤的基礎。除馬、驢、兔外，其他家畜的絨毛膜均有部分與羊膜接觸，形成羊膜絨毛膜。

（5）臍帶。臍帶是胎兒和胎盤連繫的紐帶，被覆羊膜和尿膜。臍帶隨胚胎的發育逐漸變長，使胚體可在羊膜腔中自由移動。臍帶的長度種間差異較大，

(2) Amnion

Amnion is a transparent membrane wrapped outside the fetus, which connects with the fetal skin at the umbilical perforation. The amnion is closed as the amniotic cavity, which contains amniotic fluid. The fetus floats in the amniotic fluid which acting as a buffer.

(3) Allantois

Allantois is closed as an allantoic sac, which connects with the fetal bladder through the urachus and contains urine, is equivalent to the temporary bladder outside the embryo. There are a large number of vessels from umbilical artery and umbilical vein on the allantois. The allantoic sac of pigs, cattle and sheep are surrounded by amniotic sac on the ventral side and sides of the fetus, and the allantoic sac of horses are wrapped entirely in the amniotic sac.

(4) Chorion

Chorion is the outermost layer of the embryo, which is covered with villi and is embedded in the glandular fossa of the uterine mucosa to form the basis of the fetal placenta. Apart from horses, donkeys and rabbits, the chorion of other livestock contacts with the amnion to form the amniochorion.

(5) Umbilical cord

Umbilical cord is the link between fetus and placenta, covered with amnion and allantois. The umbilical cord gradually lengthens with the development of the

豬平均為 20～25 cm，牛為 30～40 cm，羊為 7～12 cm，分娩時多數自行斷裂。

2. 胎盤

胎盤是由尿膜絨毛膜和妊娠子宮黏膜結合在一起的組織，其中尿膜絨毛膜部分稱為胎兒胎盤，而子宮黏膜部分稱為母體胎盤。

（1）胎盤的類型。根據絨毛膜表面絨毛的分布將胎盤分為瀰散型、子葉型、帶狀和盤狀 4 種類型（圖 5-5）。

embryo, so that the embryo body can move freely in the amniotic cavity. The length of the umbilical cord varies greatly in various species. The average length of the pig is 20-25 cm, the cattle is 30-40 cm, and the sheep is 7-12 cm.

2.4.2 Placenta

The placenta is a tissue that combines the urinary chorion with the pregnant uterine mucosa. The part of the urinary chorion is called the fetal placenta, and the part of the uterine mucosa is called the maternal placenta.

（1）Types of placenta

According to the distribution of villi on the chorionic surface, the placenta can be divided into four types: diffuse placenta, cotyledonary placenta, zonary placenta, discoid placenta (Figure 5-5).

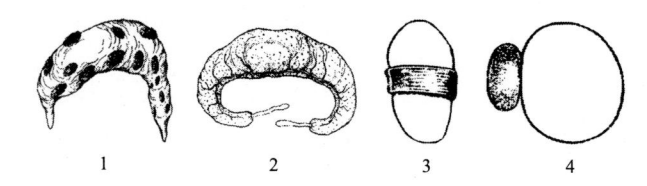

圖 5-5 胎盤的類型
Figure 5-5 Types of placenta
1. 瀰散型胎盤 2. 子葉型胎盤 3. 帶狀胎盤 4. 盤狀胎盤
1. diffuse placenta 2. cotyledonary placenta 3. zonary placenta 4. discoid placenta

①瀰散型胎盤。胎盤絨毛膜上的絨毛分散而均勻地分布在整個絨毛膜表面，如馬、驢、豬等。子宮黏膜上皮形成陷窩，絨毛深入到陷窩內。此類胎盤構造簡單，結合不牢，易發生流產；分娩時出血較少，胎衣易於脫落。

②子葉型胎盤。以反芻動物牛、羊為代表。尿膜絨毛膜上的絨毛呈叢狀分布，相應嵌入母體子宮黏膜上皮的陷窩中。此類胎盤結合緊密，產後易出現胎衣不下。

①Diffuse placenta

The villi of placenta are distributed evenly on the surface of the entire chorionic surface, such as horses, donkeys and pigs. The epithelium of the uterine mucosa forms a lacuna, and the villi penetrate into the lacuna. This type of placenta is simple in structure, weak in combination, and prone to miscarriage, less bleeding during delivery, and easy to fall off the placenta.

②Cotyledonary placenta

Cotyledonary placenta is typically seen in ruminant animals such as cattle and sheep. The villus on the urinary chorion are clustered, and correspondingly embedded in the lacuna of the epithelium of ma-

③帶狀胎盤。絨毛集中於絨毛膜的中央，呈環帶狀，故稱帶狀胎盤。犬、貓等食肉動物為此類胎盤。

④盤狀胎盤。在胎兒發育過程中，絨毛集中於一個圓形的區域，呈圓盤狀。人和靈長類屬此類型。

（2）胎盤的功能。胎盤是一種功能複雜的器官，是母體和胎兒相連接的紐帶，具有物質運輸、合成及分解代謝、分泌激素及免疫等多種功能。

（五）妊娠生理

母畜妊娠期間，由於卵巢上妊娠黃體的存在和孕體的發育，內分泌系統出現明顯的變化，使母畜的生殖器官和整個機體都出現了特殊變化，這些變化對妊娠診斷有很好的參考價值。

1. 全身的變化

母畜妊娠後，食慾增加，體重增加，被毛光潤，性情溫順。妊娠中後期，由於胎兒增長迅速，儘管母畜食慾增強，但仍入不抵出，膘情有所下降。妊娠末期，胎兒對鈣、磷等礦物質需要量增多，若不及時補充，母畜容易出現後肢跛行，牙齒磨損較快。妊娠中後期（馬、牛5個月，羊3個月，豬2個月以後），孕畜腹部膨大，排糞（尿）次數增多，行動謹慎，容易疲勞和出汗。

Project V Pregnancy Diagnosis

ternal uterine mucosa. This kind of placenta is tightly combined, and the placenta is prone to retain after delivery.

③Zonary placenta

The villi concentrate in the center of the chorion and in a shape of ring-like, so this kind of placenta is called zonary placenta, typical examples including those of dogs, cats and other predators.

④Discoid placenta

During the development of the fetus, the villi of the placenta concentrate in a circular area, which is discoid, such as those of human beings and primates.

（2）Function of placenta

The placenta is a organ with complex function and serves as the link between the mother and fetus. It has many functions, such as material transportation, synthesis, catabolism, hormone secretion and immunity.

2.5 Physiology of pregnancy

During pregnancy, the endocrine system changes obviously due to the existence of corpus luteum and the development of conceptus. Special changes have taken place in the reproductive organs and the whole body of the female animals. These changes provide a good reference value for the pregnancy diagnosis.

2.5.1 Changes of the body

After pregnancy, appetite and weight increase, fur coat is smooth and the temperament of the dam becomes gentle. In mid-late pregnancy, due to the rapid growth of the fetus, although the dam eat more, but that is not enough for the nutrient supply, so that the fatness often declines. At the end of pregnancy, the fetus needs more minerals such as calcium and phosphorus. If not supplemented in time, health problems such as limping hind limps, fast teeth wearing could appear. In mid-late pregnancy(5 months for horse and cattle, 3 months for sheep, 2 months for pigs), maternal abdomen distends, the frequency of excretion and urine increases, the movement is cautious, and easy to fatigue and sweating.

2. 生殖器官的變化

（1）卵巢。母畜妊娠後，卵巢上的妊娠黃體質地較硬，比週期黃體略大，持續存在於整個妊娠期，分泌較多孕酮，以維持妊娠。妊娠早期，卵巢偶有卵泡發育，致使孕後發情，但多數不排卵而退化、閉鎖。隨著胎兒體積增大，卵巢隨子宮沉入腹腔。

（2）子宮。妊娠期間，子宮通過增生、生長和擴展的方式以適應胎兒生長的需要。妊娠前半期，子宮體積的增長主要是子宮肌纖維增生、肥大，後半期由於胎兒生長和胎水增多，子宮壁擴張變薄。由於子宮重量增加，並向前下方垂入，因此至妊娠的後半期，一部分子宮被拉入腹腔，但至妊娠末期，由於胎兒增大，又會被推回至骨盆腔前緣。

子宮頸在妊娠期間收縮緊閉。子宮頸內腺體數目增加，分泌的黏液濃稠，充塞在頸管內形成栓塞，稱為子宮栓。子宮栓可防止外界的異物和微生物進入子宮，有保胎作用。牛的子宮頸分泌物較多，妊娠期間有子宮栓更新現象，馬、驢的子宮栓較少。子宮栓在分娩前液化排出。

（3）子宮動脈。

妊娠期內子宮血管變粗，動脈內膜增厚，且與動脈的肌層連繫變疏鬆，血液流動時出現的脈搏由原來清楚的跳動

2.5.2 Changes of reproductive organs

(1) Ovary

After pregnancy, the corpus luteum of pregnancy on the ovary is hard, slightly larger than the corpus luteum of the cycle, which persists throughout the pregnancy period and secretes more progesterone to maintain pregnancy. In the early pregnancy, follicles are occasionally developed in ovaries, resulting in estrus after pregnancy, however, most of them do not ovulate but degenerating into atresia. With the increase of fetal size, ovaries sink into abdominal cavity with uterus.

(2) Uterus

During pregnancy, the uterus adapts to the fetal growth by proliferating, growing and expanding. During the first half of pregnancy, the increase of uterus is mainly due to the hypertrophy of uterine myofibrils. In the second half of pregnancy, uterine wall will dilate and thin due to the fetal growth and increased fetal fluid. As the weight of the uterus increases and falls forward and downward, some of the uterus is pulled into the abdominal cavity in the second half of pregnancy, but at the end of pregnancy, due to the fetal enlargement, it is pushed back to the front of the pelvic cavity.

The cervix is constricted during pregnancy. The number of glands in the cervix increases, and the secretion of mucus is thick, which fills the cervical canal and causes embolism, known as uterine embolism. Uterine embolism can prevent miscarriage by preventing foreign matter and microorganisms from entering the uterus. There are more cervical secretions in cattle, with uterine embolism renewal during pregnancy, but there are fewer uterine embolism in horses and donkeys. Uterine embolism is liquefied and discharged before delivery.

(3) Uterine artery

During pregnancy, uterine blood vessels and endarterium became thicker. The connection between endarterium and arterial muscular layer becomes looser. The pulsation of blood flow changes from the clear to the

變為間隔不明顯的流水樣顫動，稱為妊娠脈搏（孕脈）。這是妊娠的特徵之一，不同母畜孕脈的強弱及出現時間不同。至妊娠末期，牛、馬的子宮動脈粗如食指。

（4）陰道和陰門。妊娠初期，陰門收縮緊閉，陰道乾澀。妊娠末期，陰唇、陰道因水腫而柔軟，利於胎兒產出。

三、妊娠期

妊娠期是母畜妊娠全過程所經歷的時間，妊娠期長短主要受畜種、品種、年齡、胎兒、環境條件等因素的影響。各種動物的平均妊娠期見表 5-1。

flow-like with no obvious interval, which is called pregnancy pulse. This is one of the characteristics of pregnancy. The intensity and appearing times of pregnancy pulses of different female animals are different. At the end of pregnancy, the uterine artery of cattle and horses is as thick as the index finger.

(4) Vagina and vulvae

At the begin of pregnancy, the vulvae contracts tightly and the vagina is dry. At the end of pregnancy, the labium and vagina are soft due to edema, which is conducive to fetal delivery.

3 Gestational period

Gestation period is the whole process of maternal pregnancy. The length of gestation period is mainly affected by species, breed, age, fetus, environmental conditions and other factors. The table below shows the average gestational periods of various animals (Table 5-1).

表 5-1 各種動物的妊娠期（d）
Table 5-1 The average gestational periods of various animals（day）

動物 Animals	平均 Average	範圍 Range	動物 Animals	平均 Average	範圍 Range
牛 Cattle	282	276～290	馬 Horse	340	320～350
水牛 Buffalo	307	295～315	驢 Donkey	360	350～370
犛牛 Yak	255	226～289	駱駝 Camel	389	370～390
豬 Pig	114	102～140	犬 Dog	62	59～65
綿羊 Sheep	150	146～161	貓 Cat	58	55～60
山羊 Goat	152	146～161	兔 Rabbit	30	28～33

任務 1 牛的妊娠診斷
Task 1 Pregnancy Diagnosis of Cattle

任務描述

準確地對配種後的母牛進行妊娠診斷，特別是早期診斷，對提高母牛的受

Task Description

Accurate pregnancy diagnosis, especially early diagnosis, is very important to improve the conception

胎率有十分重要的意義。母牛的妊娠診斷方法有多種，如外部觀察法、超音波診斷法、直腸檢查法、陰道檢查法及實驗室診斷法等。請選擇適合的方法診斷母牛是否妊娠，如已妊娠請判斷妊娠的階段，並推算其預產期。

任務實施

一、外部觀察法

（一）準備工作
母牛 10～20 頭、聽診器。

（二）診斷方法
1. 視診

詢問、調查母牛的生產階段和發情週期情況；觀察母牛的採食情況、膘情、行為、性情及體型變化；觀察母牛的外生殖器官變化和乳房變化。

2. 觸診

在母牛妊娠後期，用手掌或拳頭在其右膝壁前方、臁部下方推動腹壁感觸胎兒的「浮動」以判斷妊娠情況和妊娠時間。

3. 聽診

妊娠6個月以後，在安靜的場所於母牛右臁部下方或膝壁內側聽取胎心音。

（三）結果判定
1. 視診

經配種的母牛，過了兩個情期仍不出現發情，則可初步確定為已妊娠。母牛妊娠後，食慾增加，膘情好轉，被毛光亮，性情溫順，行動謹慎，易離群，怕擁擠。妊娠初期，外陰收縮緊閉，有皺紋。妊娠中後期（5個月以後），腹

rate of cattle. There are many methods to diagnose pregnancy in cattle, such as visual observation, ultrasonic diagnosis, rectal palpation, vaginal examination and laboratory diagnosis. Choose the appropriate method to diagnose whether the cow is pregnant or not. If the cow is pregnant, please judge the stage of pregnancy and calculate the expected date of delivery.

Task Implementation

1 Visual Observation

1.1 Preparations
10-20 cows, stethoscopes.

1.2 Diagnostic methods
1.2.1 Visual inspection

To inquire and investigate the production stages and estrus cycle of cows, to observe the feeding status, fatness, behaviors, temperaments and changes of body shape, to observe the changes of external reproduction organs and breasts.

1.2.2 Palpation

In late pregnancy, the floating fetus can be felt to judge the state and period of pregnancy by pushing the abdominal wall with the palm or fist in front of the right knee wall and below the shank.

1.2.3 Auscultation

After 6 months of pregnancy, fetal heart beats can be heard below the right shank or inside the knee wall in a quiet place.

1.3 Result judgement
1.3.1 Visual inspection

After two estrous periods, if the mated cow still does not show estrus, which can be preliminarily determined as pregnant. After pregnancy, some physiological phenomena will appear in the cow such as increased appetite, improved fatness, good bloom, gentle temperament, cautious action, easy to leave the herd and afraid of crowding. At the beginning of pregnancy, the vulva

围、乳房增大，右侧腹壁突出，有的牛会出现腹下和四肢水肿。8 个月左右，右侧腹壁可见到胎动。此方法在妊娠中后期观察比较准确，但不能在早期做出确切诊断。

2. 触诊

根据胎儿大小和母牛肥胖程度，大约不到 5% 的母牛在妊娠 5 个月时就能感触到胎动，10%～50% 于妊娠 6 个月，70%～80% 于妊娠 7 个月，90% 以上于妊娠 9 个月。

3. 听诊

一般胎心音的频率为 100 次/min 以上，为母牛心音频率的 2 倍以上。

二、直肠检查法

直肠检查是一种隔着直肠壁检查母牛卵巢、子宫及孕体状况的妊娠诊断方法，该方法可以判断母牛是否妊娠以及妊娠所处的阶段。直肠检查法是母牛早期妊娠诊断最常用和最可靠的方法之一。可在配种后 40～60 d 判断是否妊娠，准确率达 90% 以上。

（一）准备工作

母牛站立固定，将尾巴拉向一侧，掏出宿粪，清洗外阴。检查人员将指甲剪短、磨光，穿好工作服，戴上长臂手套，清洗并涂抹润滑剂。

（二）诊断方法

检查人员站在母牛正后方，将尾巴拉向一侧，手指并拢呈锥形缓慢旋转伸入直肠，摸到子宫颈，然后手掌展平向

Project V Pregnancy Diagnosis

contracts tightly and wrinkles. In the mid-late pregnancy (after 5 months), the abdominal circumference and breast enlarge, the right abdominal wall protrudes, and some cows have edema of the hypogastrium and limbs. At about 8 months, fetal movement can be observed in the right abdominal wall. This method is more accurate in the mid-late pregnancy, but can not be used to make a definite diagnosis in the early stage.

1.3.2 Palpation

According to the size of fetus and degree of obesity of cows, fetal movement can be felt in less than 5% of the cows at 5 months of gestation, 10%-50% at 6 months, 70%-80% at 7 months and 90% at 9 months.

1.3.3 Auscultation

In general, the frequency of fetal heart beats is more than 100 times per minute, which is more than twice the number of cows.

2 Rectal palpation

Rectal palpation is a method of pregnancy diagnosis that examines the ovary, uterus and embryo of cows across the rectal wall. This method can determine whether the cows are pregnant or not and the stage of pregnancy. Rectal palpation is one of the most common and reliable methods for early pregnancy diagnosis in cows. Pregnancy can be judged 40-60 days after mating, and the accuracy rate is more than 90%.

2.1 Preparations

The cow is fixed standing and the tail is pulled to one side to remove the feces in the rectum and clean the vulva. Inspectors should cut and polish their nails, put on work clothes, put on long arm gloves and apply lubricants.

2.2 Diagnostic methods

The inspector stands behind the cow, pulls the tail to one side, slowly rotates his fingers together into the rectum in a conical shape, touches the cervix, then

前滑動，分別觸摸兩個子宮角或孕體狀況。若不確定母牛是否妊娠，可繼續向前，在子宮角尖端外側或下側尋找卵巢，觸摸卵巢有無黃體存在。

檢查項目主要包括：卵巢的位置、大小和有無黃體存在；子宮角的質地、狀態、大小和位置；胚泡的大小和位置；子葉是否出現及其大小；子宮動脈的粗細和有無妊娠脈搏等。

(三) 結果判定

1. 未妊娠特徵

青年母牛的子宮及卵巢均位於骨盆腔內。經產多次的牛，生殖器官比較大，子宮角位於骨盆入口前緣的腹腔內。兩子宮角大小相等，形狀相似，彎曲如綿羊角狀，經產牛有時右角略大於左角，弛緩、肥厚。能夠清楚地摸到子宮角間溝，子宮質地良好，有彈性和收縮性。卵巢大小及形狀視有無黃體或較大的卵泡而定。

2. 妊娠特徵

(1) 妊娠 18～25 d。子宮角變化不明顯，一側卵巢上有黃體存在，可初步診斷為妊娠。

(2) 妊娠 30 d。兩側子宮角不對稱，孕角比空角粗大、鬆軟，有波動感，收縮反應不明顯，空角較厚且有彈性（圖 5-6）。

(3) 妊娠 45～60 d。子宮角和卵巢垂入腹腔，孕角比空角約粗兩倍，壁薄柔軟，波動明顯（圖 5-7）。用手

slides the palm forward and touches the two uterine horns or the embryo. If you are not sure the cow is pregnant, you can move your hand further forward to look for the ovary at the outside or below the tip of uterine horn and feel the presence or absence of the corpus luteum.

Examination items include: the location, size and presence of corpus luteum; the texture, state, size and location of uterine horn; the size and location of blastocyst; cotyledons and its size; the size and pulse of uterine artery.

2.3 Result judgement

2.3.1 Unpregnant characteristics

The uterus and ovaries of young cows are located in the pelvic cavity. In the cows with multiple births, the reproductive organs are larger, and uterine horns are in the abdominal cavity at the front edge of the pelvic entrance. The two uterine horns are equal in size and similar in shape, curved like sheep horns. Sometimes the right uterine horn is slightly larger than the left one, which is relaxed and hypertrophic. The sulcus between uterine horns can be clearly touched, the uterus is good, elastic and contractile. The size and shape of ovaries depend on the presence or absence of corpus luteum or larger follicles.

2.3.2 Pregnant characteristics

(1) 18th to 25th day of gestation

The change of uterine horn is not obvious. There is corpus luteum on one side of the ovary, which could be initially diagnosed as pregnancy.

(2) 30th day of gestation

The uterine horns are asymmetric. The gestational uterine horn is thicker and softer than the empty uterine horn. It has a sense of fluctuation and the contraction reaction is not obvious. The empty uterine horn is thicker and elastic (Figure 5-6).

(3) 45th to 60th day of gestation

The uterine horns and ovaries fall into the abdominal cavity. The gestational uterine horn is about twice

指從子宮角尖端向基部輕輕滑動，可感到有胎囊滑過，胎兒大如鴨蛋或鵝蛋。角間溝稍平坦，但仍能分辨。此時一般可確診。

as thick as that of the empty uterine horn, the wall is thin and soft with obvious fluctuations (Figure 5-7). Sliding gently from the tip of the uterine horn to the base with your finger, you can feel the fetal sac slipping through. The fetus is as big as duck's or goose's egg. The sulcus between uterine horns is slightly flat, but it can still be distinguished. At this time, the diagnosis can generally be confirmed.

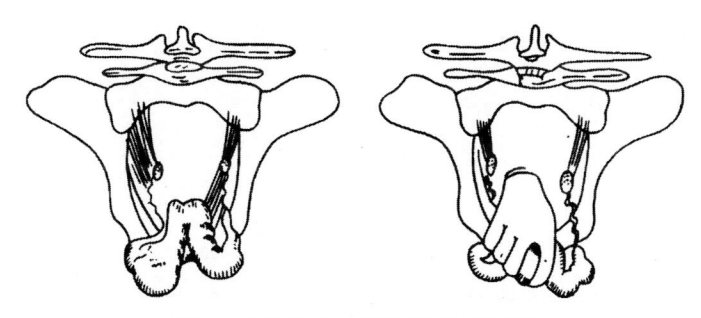

圖 5-6　牛妊娠 30 d 子宮形狀及觸診方法

Figure 5-6　Uterine shape and palpation method of 30th day of gestation in cattle

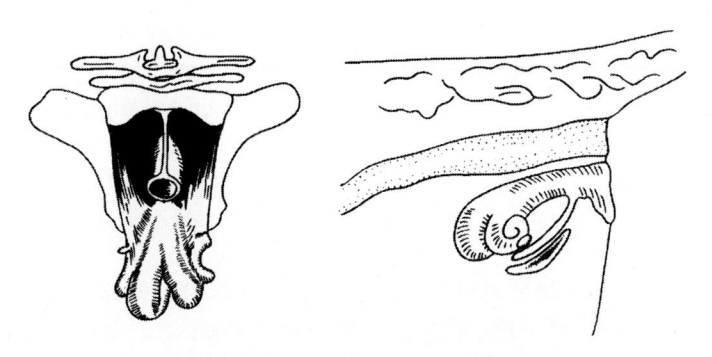

圖 5-7　牛妊娠 60 d 子宮形狀及位置

Figure 5-7　Uterine shape and location of 60th day of gestation in cattle

（4）妊娠 90 d。角間溝消失，子宮頸移至恥骨前緣（圖 5-8）。孕角大如嬰兒頭，波動明顯，空角比平時增大一倍，子葉如蠶豆大小。

(4) 90th day of gestation

The sulcus between uterine horns will disappear and the cervix will move to the front of the pubic bone (Figure 5-8). The gestational uterine horn is as big as the baby's head, and the fluctuation is obvious. The empty uterine horn is twice as big as usual, and the cotyledons are as big as broad beans.

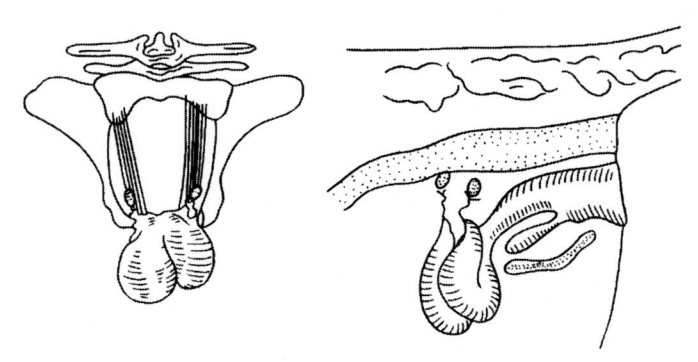

圖 5-8　牛妊娠 90 d 子宮形狀及位置
Figure 5-8　Uterine shape and location of 90th day of gestation in cattle

(5) 妊娠 120 d。子宮沉入腹底，只能觸摸到子宮後部及子宮壁上的子葉，子葉直徑為 2～5 cm。子宮頸沉移至恥骨前緣下方，不易摸到胎兒。子宮中動脈逐漸變得粗如手指，並出現明顯的妊娠脈搏。

(6) 妊娠 5 個月。子宮全部沉入腹腔，在恥骨前緣稍下方可以摸到子宮頸。子葉逐漸增大，大小如胡桃或雞蛋。孕角側子宮中動脈已明顯，空角側尚無或稍有妊娠脈搏。

(7) 妊娠 6 個月。胎兒已經很大，子宮沉至腹腔底。胎盤突如鴿蛋大小，孕角側子宮中動脈粗大，孕脈亦明顯。

(8) 妊娠 7 個月。由於胎兒更大，故從此以後都容易摸到。兩側子宮中動脈均有明顯的孕脈，但空角側較弱。

(9) 妊娠 8 個月。子宮頸回到骨盆

(5) 120th day of gestation

The uterus sinks into the bottom of the abdomen. We can only touch the the back of uterus and the cotyledons on the wall of uterus, and the diameter of the cotyledons is 2-5 cm. The cervix sinks beneath the front edge of the pubic bone, it is difficult to touch the fetus. The middle uterine artery gradually becomes thicker as a finger, and pregnancy pulses become obviously.

(6) 5th month of gestation

The whole uterus sinks into abdominal cavity, and the cervix can be touched slightly below the front edge of the pubic bone. Cotyledons increase gradually as big as walnuts or eggs. The middle uterine artery on the gestational horn is obvious, but there was no or slight pulse on the empty horn.

(7) 6th month of gestation

The fetus is very large and the uterus sinks into the bottom of the abdominal cavity. The size of placenta protuberant is like a pigeon's egg. The middle uterine artery on the side of gestational horn is thick, and the pregnancy pulses are also obvious.

(8) 7th month of gestation

Because the fetus is bigger, so it is easy to be touched thereafter. Both sides of the middle uterine artery have obvious pregnancy pulses, but the empty horn side is weak.

(9) 8th month of gestation

前緣或骨盆腔內，很容易觸及胎兒，胎盤突大如鴨蛋，兩側子宮中動脈孕脈顯著，孕角側子宮後動脈的孕脈也已清楚，個別牛即使到產前也不顯著。

（10）妊娠 9 個月。胎兒的前置部分進入骨盆入口，所有的子宮動脈均有顯著孕脈，手伸入肛門，只要貼在骨盆側壁上即可感到孕脈顫動。

（四）注意事項

在直腸檢查過程中，檢查人員應小心謹慎，避免粗暴。如遇母牛努責，應暫時停止操作，等待直腸收縮緩解時再進行檢查。母牛做早期妊娠檢查時，要抓住典型徵狀，根據子宮角和卵巢的變化，做出綜合判斷。以下幾種情況要注意區分：

（1）一些母牛配種後 20d 已妊娠，但偶爾出現假發情現象，直腸檢查妊娠徵狀不明顯，對這種牛應慎重對待，無成熟卵泡者不應配種。

（2）懷雙胎母牛，妊娠 2 個月時兩側子宮角是對稱的，不能依其對稱而判為未孕。

（3）正確區分妊娠子宮和子宮疾病。妊娠 90～120 d 的子宮容易與子宮積液、積膿混淆。子宮積液或積膿時，一側子宮角及子宮體膨大，子宮有不同程度的下沉，但子宮並無妊娠徵狀，也無子葉出現。

The cervix returns to the pelvic front or pelvic cavity, and it is easy to touch the fetus. The placenta protuberant is as big as duck's egg. The pregnancy pulses of the middle uterine artery on both sides are significant. And the pulses of the posterior uterine artery on the gestational horn are also clear. Even before delivery, the pregnancy pulses of some individual cattle are not significant.

(10) 9th month of gestation

The anterior part of the fetus enters the pelvic entrance, and all uterine arteries have significant pregnancy pulses. The pregnancy pulses can be felt when sliding the hand into the anus and attachs it to the pelvic lateral wall.

2.4 Cautions

In the process of rectal palpation, the inspectors should be careful and mild. In case of cow abdomen contracting, the operation should be suspended until rectal contraction is relieved. When cows are examined for early pregnancy, we should grasp the typical symptoms and make a comprehensive judgement according to the changes of uterine horns and ovaries. The following situations should be distinguished:

(1) Some pregnant cows occasionally appear false estrus phenomenon 20 days after mating, and pregnancy symptoms are not obvious by rectal palpation. This kind of cows should be treated carefully and should not be mated without mature follicle.

(2) Cows are pregnant with twins, the uterine horns are symmetrical at 2nd month of gestation and can not be judged infertile according to their symmetry.

(3) Correctly distingwish between pregnancy uterus and uterine diseases. The uterus of 90th-120th day of gestation is easily confused with effusion and pyometra. When uterine effusion or pyometra occurs, the horns and bodies of the uterus on one side are enlarged and the uterus sinks to varying degrees. However, the uterus has no pregnancy signs and cotyledons.

(4) 正確區分妊娠子宮與充滿尿液的膀胱。妊娠 60～90 d 的子宮，可能與充滿尿液的膀胱混淆，特別是妊娠 2 個月的子宮。膀胱輪廓清楚，兩側沒有牽連物，牛的子宮前有角間溝，後有子宮頸。膨大的膀胱表面不光滑，有網狀感。胎泡表面光滑，質地均勻。

(5) 正確區分孕脈與一些顫動脈搏。經產多次的空懷牛，子宮動脈也有類似孕脈的顫動，應加以注意。

三、超音波診斷法

超音波診斷是利用超音波的物理特性探知母牛的妊娠情況。現在生產上多採用便攜式獸用超音波儀（圖 5-9），可對母牛做出早期妊娠診斷，操作方法簡單，準確率高。

(4) Correct distinction between pregnant uterus and urinary bladder. The uterus at 60th-90th day of gestation may be confused with urine-filled bladder, especially at 2nd month of gestation. The bladder is clear, without involvement on both sides, but the cattle have a sulcus at the front of the uterus and a cervix at the back. The enlarged bladder surface is not smooth and has a reticular feeling. The surface of the fetus is smooth and is of even texture.

(5) Correct distinction between pregnancy pulses and fibrillating pulses. The uterine artery of the nonpregnant cattle with a history of multiple delivery also has tremor similar to that of the pregnancy pulses, which should be paid attention to.

3 Ultrasound diagnosis

Ultrasound diagnosis can detect the pregnancy of cows using the physical characteristics of ultrasound. At present, portable B-mode ultrasonograph (Figure 5-9) is widely used. It can diagnose early pregnancy of cows, with simple operation and high accuracy.

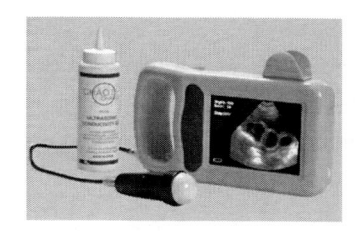

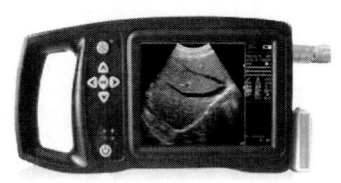

圖 5-9 獸用超音波診斷儀
Figure 5-9 Veterinary B-mode ultrasonograph

(一) 準備工作

母牛站立固定，將尾巴拉向一側，清除宿糞。檢查人員將指甲剪短、磨光，穿好工作服，挽起衣袖，戴上一次性長臂手套，塗抹滑潤劑。將防水探頭連接到超音波儀上，打開主機，並將超音波儀挎戴到手臂上。

3.1 Preparations

The cow is fixed standing and the tail is pulled to one side to remove the feces in the rectum. Inspectors cut and polish their nails, put on work clothes, roll up sleeves, put on disposable gloves and apply lubricants. Connect the water-proof probe to the machine, turn on the power and strap B-mode ultrasonograph onto your arm.

專案五 妊娠診斷
Project V Pregnancy Diagnosis

（二）檢查方法

牛的超音波儀診斷法包括體外探查法、陰道探查法和直腸探查法，其中直腸探查法應用效果最好。將塗抹耦合劑的探頭慢慢送入受檢母牛直腸，隔著直腸壁緊貼子宮角緩慢移動並不斷調整探查角度，觀察超音波實時圖像，直至出現滿意圖像為止，按凍結鍵，根據圖像判斷是否妊娠（圖 5-10）。探頭離子宮越近，探測的圖像就越清楚。

3.2 Diagnostic methods

B-mode ultrasound diagnosis in cattle includes in vitro exploration, vaginal exploration and rectal exploration, among which rectal exploration is the best one. The probe smeared with the couplant is slowly fed into the rectum of the cow, moved slowly across the rectal wall close to the uterine horns to adjust the exploration angle. The real-time image of B-mode ultrasonography is observed until satisfactory images appear. Pregnancy is judged according to the image by pressing the freezing key (Figure 5-10). The closer the probe is from the uterus, the clearer the image will be.

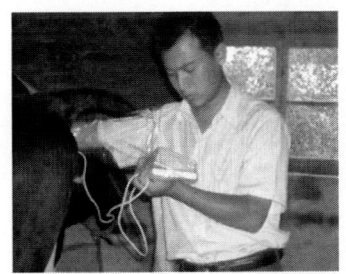

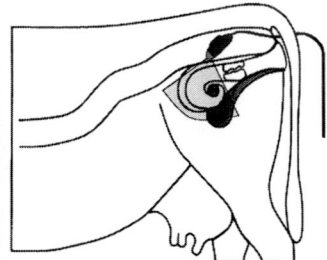

圖 5-10 牛超音波妊娠檢查
Figure 5-10 B-mode ultrasound pregnancy examination in cattle

（三）結果判定

在子宮內檢測到胚囊、胚斑和胎心搏動即判為陽性；聲像圖顯示一個或多個圓形液性暗區，判為可疑；聲像圖顯示子宮壁無明顯增厚變化、無回音暗區，判為陰性。

1. 空懷母牛子宮聲像圖

子宮體輪廓清晰，內部呈均勻的等強度回音，子宮壁很薄（圖 5-11）。

2. 妊娠母牛子宮聲像圖

妊娠母牛的子宮壁增厚，配種後 12～14d 子宮腔內出現不連續、無反射小區，即為聚有液體的胚泡。妊娠 20d 胚泡結構中出現短直線狀的胚體。妊娠

3.3 Result judgement

Embryo sac, plaque and fetal heart beat are detected in uterus, which could be judged as positive. The sonogram shows one or more circular liquid dark areas, which is suspected. The sonogram shows that there is no obvious change in uterine wall and no echo dark areas, which is negative.

3.3.1 Uterine sonogram of nonpregnant cow

The uterus has a clear outline, uniform internal echo of equal intensity, and a thin uterine wall (Figure 5-11).

3.3.2 Uterine sonogram of pregnant cow

The uterine wall of pregnant cows become thicker, discontinuous non-reflex zones appear in uterine cavity 12th to 14th day after mating, they are liquid-filled blastocysts. A short and straight linear embryo appears in the blastocyst at 20th day of gesta-

22d 可探測到胚體心跳。妊娠 22～30d，胚體呈 C 形（圖 5-12）。33～36d，呈現清晰的胚囊和胚斑圖像，胚囊如一指大小，胚斑如 1/3 指大小（圖 5-13），子宮壁結構完整，邊界清晰，胚囊液性暗區大而明顯。40d 以上，胚囊和胚斑均明顯可見，有時還可見胎心搏動（圖 5-14）。

tion. The heartbeat of embryo could be detected at 22nd day of gestation. The shape of embryo is like letter C at 22nd-30th day of gestation (Figure 5-12). The embryo sac and plaque could be clearly displayed at 33rd-36th day of gestation, the embryo sac is the size of one finger, and the plaque is the size of one third finger (Figure 5-13). The structure of uterine wall is complete, the boundary is clear, and the liquid dark area of embryo sac is large and obvious. Over 40 days, the embryo sac and plaque are clear, and sometimes fetal heartbeat is also visible (Figure 5-14).

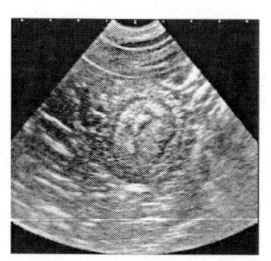

圖 5-11　未妊娠牛子宮
Figure 5-11　Uterus of nonpregnant cow

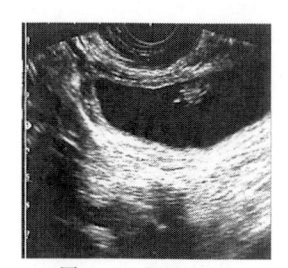

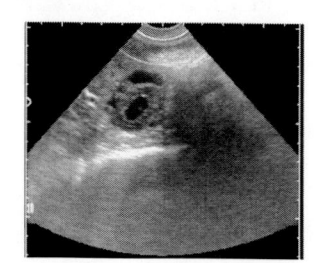

圖 5-12　妊娠 27 天　　　　　　圖 5-13　妊娠 33 天
Figure 5-12　27th day of gestation　　Figure 5-13　33rd day of gestation

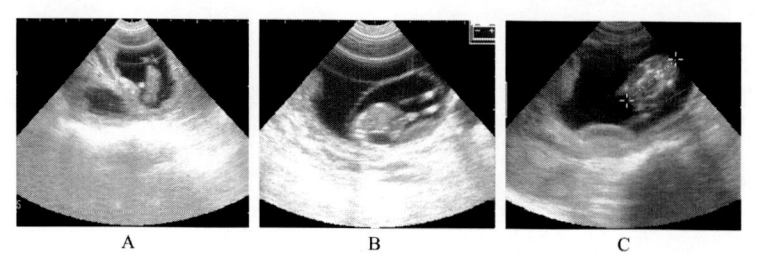

圖 5-14　妊娠 40d 以上
Figure 5-14　Pregnancy over 40 days
A. 妊娠第 45 天　B. 妊娠第 66 天　C. 妊娠第 89 天
A. 45th day of gestation　B. 66th day of gestation　C. 89th day of gestation

（四）注意事項

（1）液性暗區是圓形或近圓形強回音區（一般為純黑色），在暗區周圍有一圈完整的灰色環狀組織。

（2）暗區一定位於子宮角內部。部分假象圖是一些擠壓空腔或者血管形成，位於子宮角外面，周圍沒有完整灰色環狀組織。探查時還有可能誤將黃體或大卵泡當作胚囊。

（3）當牛患有子宮炎或子宮積液等疾病時，圖像也會出現許多不規則黑色暗區或暗區內有光斑（暗區也位於子宮角內），從而導致誤診。

3.4 Cautions

(1) The liquid dark area is a circular or near-circular strong echo area(usually pure black), surrounded by a circle of intact gray annular tissue.

(2) Dark areas must be located inside the uterine horns. Some dark areas show squeezing cavity or blood vessels, located outside the uterine horns, and there is no intact gray ring around them. It is also possible to mistake corpus luteum or large follicles as embryo sacs during exploration.

(3) When cows suffer from hysteritis or uterine hydrops, there are also many irregular black areas or dark areas with spots. The dark areas are also located in the uterine horn, which sometimes leads to misdiagnosis.

任務 2　羊的妊娠診斷
Task 2　Pregnancy Diagnosis of Sheep

任務描述

隨著養羊企業的快速發展，需要大量繁殖優質羔羊擴大生產規模。因此，母羊配種後應儘早進行妊娠診斷，及時做好空懷母羊的補配工作。如何快速辨別母羊是否妊娠？

Task Description

With the rapid development of sheep breeding enterprises, a large number of high-quality lambs are required to expand production scale. Therefore, the pregnancy diagnosis should be carried out as soon as possible after mating and do the breeding work of the non-pregnant ewes in time. How to quickly identify whether the ewe is pregnant or not?

任務實施

一、外部觀察法

1. 準備工作

將母羊放於運動場內，讓其自由活動。

2. 診斷方法

詢問、調查母羊的生產階段和發情週期情況；觀察母羊的採食情況、膘

Task Implementation

1　Visual observation

1.1　Preparations

Put the ewes in the playground and let them move freely.

1.2　Diagnostic methods

To investigate the production stage and estrus cycle of ewes. To observe the feeding status, fatness, behavior,

情、行為、性情及體型變化；觀察母羊的外生殖器官變化和乳房變化。

3. 結果判定

母羊妊娠後，發情週期停止，食慾增加，毛色光亮，性情溫順，行動謹慎小心、好靜、喜臥。妊娠初期，外陰收縮緊閉，有皺紋。妊娠2～3個月，腹圍增大，右後腹部突出，乳房增大。

二、觸診法

（一）一般觸診法

檢查人員面對母羊頭部，雙腿夾持母羊頸部，兩手以抬抱的方式在母羊腹壁前後滑動，觸摸是否有胎泡。

（二）直腸-腹壁觸診法

1. 診斷方法

將待查母羊用肥皂灌洗直腸排出糞便，使其仰臥，將塗抹潤滑劑的觸診棒（直徑1.5cm、長50cm）貼近脊椎插入直腸約30cm。一隻手用觸診棒把直腸輕輕挑起以便托起胎泡，另一隻手在腹壁上觸摸。

2. 結果判定

觸診時，如有胎泡即表明已妊娠；如果摸到觸診棒，將棒稍微移動位置，反覆挑起觸摸2～3次，仍摸到觸診棒即表明未妊娠。使用該方法時，動作要小心、輕緩，以防損傷直腸和胎兒。該方法準確率很高，在早期妊娠診斷非常重要。

三、超音波探測法

（一）準備工作

1. 固定母羊

用超音波診斷儀探查早期妊娠時，母羊側臥、仰臥或站立均可。如大群檢

---

temperament and changes of body shape. To observe the changes of external reproduction organs and breasts.

1.3 Result judgement

After pregnancy, the estrus cycle of ewes stops, some physiological phenomena appear, such as increased appetite, good bloom, gentle temperament, cautious action. At the beginning of pregnancy, the vulva contractes tightly and wrinkles. At 2nd-3rd month of gestation, the abdominal circumference and breast enlarge, the right abdominal wall protrudes.

2 Palpation

2.1 General palpation

The inspector holds the ewe's neck with both legs facing the head, and slide his hands around the abdominal wall to determine whether there are embryos.

2.2 Palpation of rectum-abdominal wall

2.2.1 Diagnostic methods

The ewe undergoes a rectal lavage with soap water to excrete feces from the rectum and made to lie on its backs. The palpation stick(1.5 cm in diameter and 50 cm in length) with lubricant is inserted into the rectum close to the spine for about 30 cm. One hand gently lifts the rectum with a palpation stick to hold up the fetus, and the other hand feels the abdominal wall.

2.2.2 Verdict-making

When palpating, if there is a fetus, it indicates pregnant. If touching the palpation stick, you need to move the stick slightly, repeatedly lift and touch 2-3 times, it indicates unpregnant if still touching the palpation stick. We should be careful and gentle to prevent damage to the rectum and fetus. The accuracy of this method is very high, and it is very important in early pregnancy diagnosis.

3 Ultrasound diagnosis

3.1 Preparations

3.1.1 Fixed ewes

B-mode ultrasonograph can detect early pregnancy of ewes by lying on the side or back or standing. For

專案五　妊娠診斷

Project V　Pregnancy Diagnosis

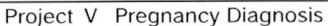

查，採用固定架固定；如少量檢查，可由助手扶持，使母羊安靜站立。如需進行腹側檢查，則需在右腹側後部剪毛。

2. 超音波儀準備

連接好超音波儀探頭，打開超音波儀，調節好對比度、輝度和增益，使其適合當時當地的光線強弱及檢測者的視覺。準備好耦合劑。

（二）診斷方法

1. 直腸探查

應用於妊娠早期（妊娠 40 d 以內），將探頭插入直腸內約 15 cm，越過膀胱，向兩側轉 45°角進行掃查，以探測到胎水或子葉為判定妊娠陽性依據。

2. 腹部探查

檢查者蹲於羊體一側，將塗抹耦合劑的探頭緊貼母羊腹部皮膚，朝盆腔入口方向定點做扇形掃查，如探查到胎兒（胎頭、胎心、脊椎或胎蹄）可判定為陽性。妊娠早期在乳房兩側及膝皺襞之間無毛區域，或兩乳房的間隔處進行探查；妊娠中後期可在右側腹壁進行探查。

（三）結果判定

1. 未妊娠母羊

子宮一般位於膀胱前方或前下方，為近圓形的弱反射，直徑在 10.0 mm 以上。有時斷面中央可見一個很小的暗區，直徑為 2.0~3.0 mm，但不隨配種天數增加而變大，子宮可隨膀胱積尿的程度而上下移位。

mass inspection, the sheep should be fixed by a frame. For individual inspection, the ewe can be supported by an assistant to make them stand quietly. If ventral examination is needed, the wool of back part of right ventral need to be shorn.

3.1.2　Preparation of B-mode ultrasonograph

Connect the probe, turn on the B-mode ultrasonograph, adjust the contrast, brightness and gain so as to make it suitable for the local light intensity and the inspector's vision at that time. Prepare the couplant.

3.2　Diagnostic methods

3.2.1　Rectal exploration

Rectal exploration is applied in the early pregnancy (within 40 days). The probe is inserted into the rectum about 15 cm deep, crossed the bladder and turned 45 degrees to both sides for scanning. Pregnancy can be determined if fetal fluid or cotyledons are detected.

3.2.2　Abdominal exploration

The examiners squat on one side of the sheep, press the probe with the coupling agent close to the abdominal skin, and make a fixed-point sector scanning towards the pelvic entrance. If the fetus (head, heart, spine or hoof) is detected, it can be judged to be pregnant. In the early pregnancy, the exploration site is located at hairless area between the breast and the knee fold, or the interval between the two breasts. In the middle and late pregnancy, the exploration site is located at the right abdominal wall.

3.3　Result judgement

3.3.1　Unpregnant ewes

The uterus is usually located in front of or below the bladder. It is a nearly circular weak reflex with a diameter of more than 10.0 mm. Sometimes a small dark area with a diameter of 2.0-3.0 mm can be seen in the center of the section, but it does not become larger as time passes after mating. The uterus can move up and down with the variations of bladder urine amout.

2. 妊娠母羊

子宮內出現暗區（胚囊），最初為單個小暗區。配種後18～20d，可探查到胚斑，為橢圓形的弱反射光斑，長約6mm，位於子宮暗區的下方或一側。妊娠23d時，膀胱下面出現帶狀的無回音區（胎水）。妊娠25d時，呈不規則的無回音區，可見到胎體，但看不到胎心搏動。妊娠30d時，不規則無回音區擴大並移至膀胱前下方，無回音區邊緣或其中有「扣狀」小回音，為胎盤，羊膜囊內有胎體，細心觀察可見到胎心搏動。診斷早孕的準確率為97％左右。

3.3.2 Pregnant ewes

Initially a small dark area (embryo sac) appears in the uterus. Embryo spot can be detected 18-20 days after mating. It is an elliptical weak reflection spot with about 6 millimeters long, located below or on one side of the dark area of uterus. On the 23rd day of gestation, a banded anechoic zone (fetal fluid) appears below the bladder. On the 25th day of gestation, there is an irregular anechoic zone in which the fetal body could be seen, but no fetal heart beat. On the 30th day of gestation, the irregular anechoic area is enlarged and moves to the anterior and inferior part of the bladder. The edge of the anechoic area or with the small 「button」 echo in it is the placenta, and the fetal heart beat can be observed carefully in the amniotic sac. The accuracy of early pregnancy diagnosis is about 97％.

任務3　豬的妊娠診斷
Task 3　Pregnancy Diagnosis of Pigs

任務描述

母豬配種後，應儘早檢出空懷母豬，及時補配，防止空懷。這對於保胎、縮短胎次間隔、提高繁殖力和經濟效益具有重要的意義。母豬妊娠診斷常用的方法有外部觀察法、返情檢查法、超音波診斷法等。規模化豬場多採用B型超音波檢測，也有用A型超音波檢測。二者有哪些異同點？

Task Description

After mating, the nonpregnant sows should be detected as soon as possible, and be mated again in time. This is of great significance to the preservation of fetus, shortening the interval of parities, improving the fecundity and economic benefits. The commonly used methods of sow pregnancy diagnosis include visual observation, re-estrus inspection and ultrasound diagnosis. Pigs in large farms are mostly detected by B-mode ultrasonograph, together with some using A-mode ultrasonograph. What are the similarities and differences between them?

任務實施

一、外部觀察法

1. 準備工作

母豬10～20頭，自由活動。

Task Implementation

1　Visual observation

1.1　Preparations

To prepare 10-20 sows and let them move freely.

Project V　Pregnancy Diagnosis

2. 診斷方法

詢問、調查母豬的生產階段和發情週期情況；觀察母豬的採食情況、膘情、行為、性情及體型變化；觀察母豬的外生殖器官變化和乳房變化。

3. 結果判定

母豬妊娠後，表現為發情週期停止，食慾增強，膘情好轉，被毛光亮，性情溫順，行動小心。妊娠初期，外陰蒼白、皺縮，妊娠中後期（2 個月後），腹圍增大、隆起。

二、 返情檢查

妊娠診斷最普通的方法是根據配種後 17～24 d 是否恢復發情。母豬的發情週期一般為 21 d 左右，正常情況下，母豬配種後 20 多天不再出現發情表現，可初步確定妊娠，第二個情期仍不發情，即可確定妊娠。個別母豬妊娠後，偶爾也會有發情表現，因此，最好與超音波診斷結合使用。

三、 超音波診斷法

（一） A 型超音波診斷母豬妊娠

1. 準備工作

待檢母豬、A 型超音波診斷儀（圖 5-15）、耦合劑。

2. 診斷方法

將待測母豬站立固定，在右腹側最後一對乳頭上方處，塗抹適量耦合劑，將探頭緊貼皮膚對子宮進行扇形掃描。當發出穩定聲音後，記錄結果。然後，在左側再檢測一次，以驗證結果。

1.2　Diagnostic methods

To investigate the production stage and estrus cycle of sows. To observe the feeding status, fatness, behavior, temperament and changes of body shape. To observe the changes of external reproduction organs and breasts.

1.3　Result judgement

After pregnancy, the estrus cycle of sows stops, some physiological phenomena appear such as increased appetite and fatness, good bloom, gentle temperament, acting cautiously. At the beginning of pregnancy, the vulva is pale and wrinkled. The abdominal circumference is enlarged and bulged in the middle and later period of pregnancy(2 months later).

2　Re-estrus inspection

The most common method of pregnancy diagnosis is based on whether estrus is restored 17-24 days after mating. The estrus cycle of sows is generally about 21 days. Normally, pregnancy is preliminarily determined if the sow does not show estrus for 20 days after mating. Pregnancy can be determined if the second cycle is still not estrus. Some sows will occasionally have estrus behaviour after pregnancy, so it is best to diagnose pregnancy in combination with ultrasonic.

3　Ultrasonic diagnosis

3.1　Pregnant diagnosis of sows by A-mode ultrasonograph

3.1.1　Preparations

Sows, A-mode ultrasonograph(Figure 5-15), coupling agent.

3.1.2　Diagnostic methods

The sow is fixed standing. Smear some coupling agent on the top area of the last pair of nipples on the right ventral side, then attach the probe to the uterus for sector scanning. When a steady sound is emitted, the results are recorded. Then, check again on the left to verify the results.

圖 5-15　豬用 A 型超音波診斷儀
Figure 5-15　A-mode ultrasonograph for swine

3. 結果判定

當聽到連續的「嘀嘀」聲，可診斷為妊娠；當聽到斷續的「嘀嘀」聲，多次調整探頭方向，仍無連續響聲，則診斷為未妊娠。

(二) B 型超音波診斷母豬妊娠

1. 準備工作

妊娠 18～35 d 的母豬數頭，超音波診斷儀，耦合劑。

2. 診斷方法

用濕毛巾擦除母豬腹部汙物。母豬站立時在探頭上塗抹耦合劑，側臥或趴臥時在探查部位塗抹耦合劑，將探頭指向恥骨前部和骨盆入口方向。隨妊娠天數的增加，探查部位逐漸前移。

3. 結果判定

(1) 妊娠母豬（圖 5-16）。當看到典型的孕囊暗區即可確認早孕陽性。妊娠早期（妊娠 18～21 d）子宮中出現孕囊，呈直徑約 1 cm 的圓形暗區，通常為一個或 2～3 個相鄰的暗區，位於膀胱暗區的前下方。隨妊娠期的延長，暗區不斷擴大且呈多個不規則圓形、橢圓形暗區。妊娠 26 d 後，胚胎逐漸顯出

3.1.3　Result judgement

Pregnancy can be diagnosed when the device emits continuous beep sound. When hearing intermittent beep sound, adjusting the direction many times, if there is still no continuous sound, unpregnancy can be diagnosed.

3.2　Pregnant diagnosis of sows by B-mode ultrasonograph

3.2.1　Preparations

Some sows at 18-35 days of gestation, B-mode ultrasonograph, coupling agent.

3.2.2　Diagnostic methods

Wipe the sow's abdomen clean with a wet towel. When the sow is standing, the probe is coated with coupling agent. While lying on the side or on the abdomen, the exploratory site is coated with coupling agent. Place the probe in the direction of the anterior part of the pubic bone and the pelvic entrance. As pregnancy progresses, the site is moved forward gradually.

3.2.3　Result judgement

(1) Pregnant sows (Figure 5-16)

When the typical dark area of the gestational sac appears, the positive early pregnancy can be confirmed. In early pregnancy (18th-21st day), there are gestational sacs in the uterus, usually one or 2-3 adjacent dark areas about 1 cm in diameter, located in the front and bottom of the bladder. As pregnancy progresses, dark areas continue to expand, showing irregular circular and oval shapes. After 26 days of

胎兒固有輪廓，胎頭、軀體及四肢逐漸發育完善，出現胎動及內臟器官（肝、胃等）。

（2）空懷母豬。空懷母豬子宮顯示為不規則圓形的弱反射區。一週後應及時複查，以免誤診。

gestation, the embryo gradually shows the inherent contour, the fetal head, body and limbs are well-developed, and fetal movement and visceral organs (liver, stomach, etc.) appear.

(2) Nonpregnant sows

The uterus of nonpregnant sows shows a weak reflex area of irregular circular shape. A week later, another examination should be made to avoid misdiagnosis.

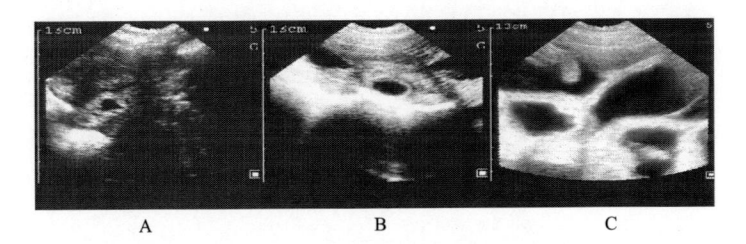

圖 5-16　豬 B 型超音波妊娠檢測聲像

Figure 5-16　Sonography of pregnant diagnosis in sows by B-mode ultrasonograph

A. 妊娠 18d　B. 妊娠 21d　C. 妊娠 28d

A. 18th day of gestation　B. 21st day of gestation　C. 28th day of gestation

專案六　接產與助產
Project Ⅵ　Delivery and Midwifery

專案導學

接產和助產是家畜繁殖中的一項重要工作，助產不當或不及時助產直接關係到母仔生命的安危及產後疾病的預防。因此，生產中既要掌握母畜接產與助產的基本理論，還要能熟練進行接產與助產操作。

Project Guidance

Delivery and midwifery are important tasks in livestock reproduction. Improper or untimely midwifery may have a direct impact on the safety of newborns and their mothers as well as postpartum diseases prevention. Therefore, we should not only grasp the basic theory of delivery and midwifery, but also be skilled in operation.

學習目標

>>> 知識目標

- 熟悉母畜的分娩預兆和分娩過程，為判斷是否正常分娩奠定基礎。
- 瞭解難產的原因和種類。
- 熟悉難產救助的基本原則。
- 理解分娩機理及影響分娩過程的因素。

>>> 技能目標

- 能準確判斷母畜的分娩預兆及難產跡象。
- 能熟練實施母畜分娩前準備及助產技術。
- 能正確掌握難產檢查及處理技術。
- 會正確護理新生仔畜。

Learning Objectives

>>> Knowledge Objectives

- Be familiar with the parturition signs and parturition process, which provides the basis for correctly diagnosing whether the parturition is normal or not.
- Know the causes and kinds of dystocia.
- Be familiar with the rules of rescue when dystocia occurs.
- Understand the mechanism and influencing factors of parturition process.

>>> Skill Objectives

- Judge the parturition sings and dystocia signs accurately.
- Master the prenatal preparation work and the skills of midwifery.
- Master the examination and treatment skills of dystocia correctly.
- Nurse newborn animals correctly.

相關知識

一、分娩機理

分娩是指雌性動物經過一定的妊娠期,胎兒在母體內發育成熟,母體將胎兒及其附屬物從子宮內排出體外的生理過程。引起分娩啟動的因素是多方面的。分娩是由激素、神經和機械等多種因素的協同、配合,母體和胎兒共同參與完成的。

1. 機械刺激

妊娠末期,由於胎兒生長很快,胎水增多,胎兒運動增強,使子宮不斷擴張,承受的壓力逐漸升高。當子宮的壓力與子宮肌高度伸張狀態達到一定程度時,便可引起神經反射性子宮收縮和子宮頸的舒張,從而導致分娩。

2. 母體激素的變化

臨近分娩時,母體內孕激素分泌減少或消失,雌激素、前列腺素($PGF_{2\alpha}$)、催產素分泌增加,同時卵巢及胎盤分泌的鬆弛素促使產道鬆弛。母體在這些激素的共同作用下發生分娩。分娩前後相關激素的變化趨勢見圖6-1。

Relevant Knowledge

1 Mechanism of parturition

Parturition is defined as the physiologic birth process by which the pregnant uterus delivers the fetus and placenta from the maternal organs. Parturition involes both the matrixes and the fetus, and it is completed by a complex interaction of many factors, such as hormonal, neural, mechanical factors, etc.

1.1 Mechanical stimulation

The fetus grow fast, motion enhancement and the fetal water increase induce the pressure on the uterus distending and expand at the final stage of gestation. When the pressure on uterus and the stretched condition of myometrium reach up to a point, it will cause nerve reflex contraction of uterus and diastole of cervix, which then leads to parturition.

1.2 Hormonal changes in dams

Near parturition, the levels of progesterone fall or disappear, while the levels of estrogen, prostaglandin ($PGF_{2\alpha}$) and oxytocin increase, at the same time relaxin secreted by ovary and placenta, gives the birth canal laxity. The variation curve of related hormones before and after parturition is shown in Figure 6-1.

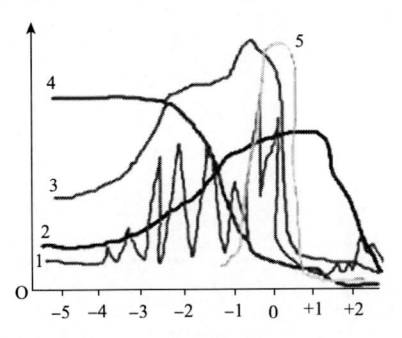

圖6-1 母畜分娩前後體內激素水準的變化

Figure 6-1 Hormonal changes in dams before and after parturition

1. $PGF_{2\alpha}$ 2. 雌激素 3. 胎兒腎上腺皮質激素 4. 孕激素 5. 催產素

1. prostaglandin ($PGF_{2\alpha}$) 2. estrogen 3. fetal adrenocorticotropic hormone 4. progesterone 5. oxytocin

3. 神經系統

神經系統對分娩並不是完全必需的，但對分娩過程具有調節作用。如胎兒的前置部分對子宮頸和陰道產生刺激，通過神經傳導使垂體後葉釋放催產素導致分娩。此外，多數母畜在夜間分娩，可能是由於外界光線及干擾減少，黑暗和安靜的環境易於接受來自子宮及軟產道的神經刺激。

4. 胎兒因素

胎兒發育成熟後，腦垂體分泌促腎上腺皮質激素，促使胎兒腎上腺分泌腎上腺皮質激素。胎兒腎上腺皮質激素引起胎盤分泌大量雌激素及母體子宮分泌大量前列腺素，並使孕激素水準下降。雌激素使子宮肌對各種刺激更加敏感，而且還能促使母體自身釋放催產素。所以在母體的催產素與前列腺素的協同作用下，激發子宮收縮，導致胎兒娩出。

5. 免疫學機理

妊娠後期，胎兒發育成熟時，胎盤發生脂肪變性，胎盤屏障受到破壞，胎兒和母體之間的連繫中斷，胎兒被母體免疫系統辨識為「異物」而排出體外。

二、 分娩預兆

母畜在分娩前，在生理和形態上都會發生一系列變化，通常將這些變化稱為分娩預兆。根據乳房、外陰部、骨盆等變化，可預測母畜的分娩時間，以便做好產前準備，確保母仔平安。

1.3 Nervous system

The nervous system is not absolutely necessary for parturition, but it regulates the process of parturition. For example, the fore-lying part of fetus can stimulate cervix and vagina, leading to posterior pituitary release oxytocin by nerve conduction, then cause labour. Otherwise, most dams labour at night probably because of the reduction of ambient light and perturbation making them easy to accept nerve stimulation from uterus and soft birth canal in dark and quiet environment.

1.4 Fetal influence

After the fetus matures, its pituitary gland secretes adrenocorticotropic hormone, induces adrenal gland to secrete adrenal hormone. Fetal contrical hormone not only causes placenta and maternal uterus to secrete large amounts of estrogen and prostaglandin, but also decreases the level of progestin. Estrogen will make the myometrium more sensitive to various stimuli and prompt the matrixes to release oxytocin. The synergistic effect of the oxytocin and prostaglandin induces the uterine contraction, then leads to the delivery of the fetus.

1.5 Immunological mechanism

When the fetus matures, it has been shown to induce placental steatosis, while the placenta barrier is destroyed and the connection between the fetus and the matrixes is interrupted at the final stage of gestation. Finally, the fetus is recognized as a foreign-body by the maternal immune system and thus being「expelled」from the body.

2 Signs of parturition

A series of physiological and morphological changes in prenatal dams are usually called signs of parturition. We can predict the parturition time of dams according to the changes of breast, vulva and pelvis, in order to make prenatal preparation and ensure the safety of both mothers and offspring.

分娩前乳房迅速發育、膨脹增大、有時還出現浮腫、個別有漏乳現象。牛的乳房變化比其他家畜明顯。在分娩前數天到一週左右，陰唇逐漸變鬆軟、腫脹、體積增大，陰唇皮膚上的皺褶展平，並充血稍變紅。從陰道流出的黏液由濃稠變稀薄，尤以牛和羊最為明顯。骨盆韌帶從分娩前1～2週開始軟化，到分娩前12～36 h，薦坐韌帶後緣變得非常鬆軟，外形消失，尾根兩側下陷，只能摸到一堆鬆軟組織，即「塌窩」，但初產牛這些變化不明顯。

分娩前精神狀態變化比較明顯，母畜出現食慾不振、精神沉鬱、徘徊不安和離群尋找安靜地等現象。豬在臨產前6～12 h，出現啣草做窩現象。家兔有扯咬胸部被毛和啣草做窩現象。馬和驢在臨產前數小時，表現不安、頻繁舉尾、蹄踢下腹部和時常起臥及回顧腹部等。

三、分娩過程

（一）影響分娩過程的因素

1. 產力

將胎兒從子宮中排出的力量稱為產力，包括陣縮和努責。陣縮是指子宮肌有節律的收縮，貫穿於整個分娩過程，是分娩的主要動力。努責是指腹肌和膈肌的強而有力收縮，是胎兒產出的輔助動力。

2. 產道

產道是分娩時胎兒由子宮排出體外的通道，可分為軟產道和硬產道兩部分。

Obvious enlargement of the mammary gland occurs before parturition, as well as swollen, and sometimes edema or milk leakage occurs, especially in cows. From a few days to a week before parturition, the labia become pliable, edematous, bulky gradually and slightly reddish with congestion, the wrinkles on the skin flatten out. Mucus from the vagina becomes thinner, especially in the cow and the ewe. The pelvic ligament becomes relaxed and flaccid about 1-2 weeks before parturition. Around 12-36 hours before parturition, the posterior border of the sacrosciatic ligament becomes more flaccid and the root of tail sags bilaterally, while only a pile of soft tissues can be touched. But these changes are not obvious in primiparous cows.

Before parturition, the mental state of the dam will change significantly, with symptoms such as anorexia, depression, restlessness and seeking isolation to seek quiet place. Swines build nests about 6-12 hours before parturition, rabbits bite the chest coat and make nests. Horses and donkeys show restlessness, frequent tail lifting, kicking the lower abdomen, frequent sitting-up and lying-down and looking back at the abdomen in few hours before parturition.

3 Parturition process

3.1 Factors affecting the parturition process

3.1.1 Force of delivery

Force of delivery means the force that expels the fetus from the uterus, including contracture and straining. Contracture means the rhythmic myometrial contractions, which is the main power of parturition and runs through the whole process of parturition. Straining refers to the strong contraction of abdominal and diaphragm muscles, which is the auxiliary power of labor.

3.1.2 Birth canal

The birth canal is the channel through which the fetus is expelled from the uterus, it can be divided into two parts: soft birth canal and hard birth canal.

(1) 軟產道。包括子宮頸、陰道、前庭和陰門。分娩時子宮頸逐漸鬆弛直至完全開張，陰道、陰道前庭和陰門也能充分鬆軟擴張。

(2) 硬產道。就是骨盆，可分為四個部分：①入口，即骨盆的腹腔面，入口大而傾斜，形狀圓而寬闊，胎兒則容易通過。②骨盆腔，即骨盆入口至出口之間的空間。③出口，即骨盆腔向臀部的開口。④骨盆軸，代表胎兒通過骨盆腔時所走的路線，骨盆軸越短越直，胎兒通過越容易。各種母畜的骨盆特點如表 6-1 和圖 6-2 所示。

3. 胎兒與母體的關係

分娩時，胎兒與母體產道的相互關係，對胎兒產出有很大影響。此外，胎兒的大小和是否畸形也影響胎兒能否順利產出。

(1) Soft birth canal

The soft birth canal includes cervix, vagina, vestibule and vulva. During parturition, the cervix gradually relaxes until it is fully opened, and the vagina, vestibule and vulva are also fully relaxed and dilated.

(2) Hard birth canal

The hard birth canal is the pelvis, which can be divided into four parts: ① The entrance is the abdominal side of the pelvis. The entrance to the pelvis is large and oblique, round and wide in shape, which is easy for the fetus to pass through. ② The pelvic cavity refers to the space between the entrance and exit of the pelvis. ③ The outlet means opening of the pelvic cavity to the buttocks. ④ The pelvic axis is the route through which the fetus passes through the pelvic cavity. The shorter and straighter the pelvic axis is, the easier for the fetus to pass through. The pelvic characteristics of various dams are shown in table 6-1 and Figure 6-2.

3.1.3 The relationship between fetus and dam

During parturition, whether the fetus can be delivered smoothly is influenced by the relationship between the fetus and the maternal birth canal, the size of the fetus and whether it is malformed.

表 6-1　各種母畜骨盆特點
Table 6-1　Pelvic characteristics of dams

	牛 Cattle	馬 Horse	豬 Swine	羊 Sheep
入口 Entrance	豎長橢圓 Vertical ellipse	圓形 Circular	近乎圓形 Almost circular	橢圓形 Ellipse
出口 Outlet	較小 Small	大 Large	很大 Great	大 Large
傾斜度 Inclination degree	較小 Small	大 Large	很大 Great	很大 Great
骨盆軸 Pelvic axis	曲線形 Curvilinear	淺弧形 Shallow arc	較直 Relatively straight	弧形 Arc
分娩難易程度 Difficulty of parturition	較難 Relatively difficult	易 Easy	很易 Very easy	易 Easy

專案六 接產與助產

Project VI Delivery and Midwifery

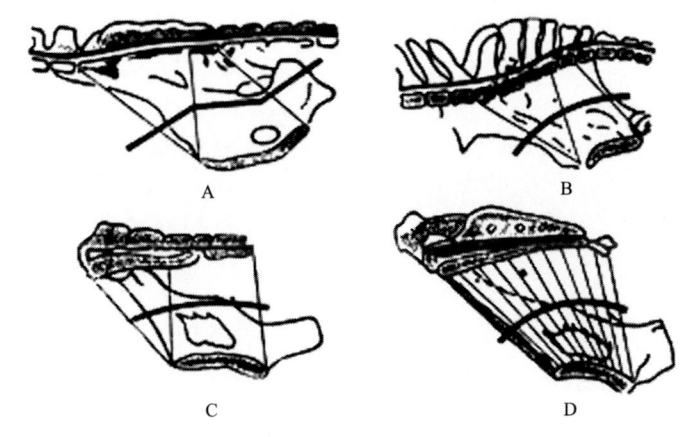

圖 6-2 各種母畜的骨盆軸
Figure 6-1 Pelvic axis of dams
A. 牛　B. 馬　C. 豬　D. 羊
A. cow　B. horse　C. sow　D. ewe

（1）胎向。指胎兒縱軸與母體縱軸的相互關係，分為縱向、豎向和橫向。縱向是正常胎向，橫向和豎向都屬反常胎向，易發生難產。

（1）Fetal orientation

Fetal orientation refers to the relationship between longitudinal axis of the fetus and that of matrixes, which can be divided into longitudinal, vertical and transverse. Normal fetal orientation is longitudinal, the others are abnormal orientations and prone to causing dystocia.

（2）胎位。指胎兒背部與母體背部的關係，分為上位、下位和側位。上位是正常的，下位和側位是反常的。如果側位傾斜不大，仍可視為正常。

（2）Fetal position

Fetal position refers to the relationship between the dorsum of the fetus and that of matrixes, which can be divided into upper, inferior and lateral position. The upper position is normal, the inferior and lateral positions are abnormal. It is still regarded as normal if inclination angle of the lateral position is not obvious.

（3）前置。又稱先露，是指胎兒先進入產道的部位。頭和前肢先進入產道為頭前置（正生）；臀部和後肢先進入產道為臀前置（倒生）。

（3）Presentation

Presentation means that the part of fetus enters the birth canal first. If the head and forelimbs first enter the birth canal, it is called anterior presentation, while if the buttocks and hind limbs first enter the birth canal, it is called posterior presentation.

（4）胎勢。指胎兒在母體內的姿勢，一般分為伸展或屈曲的姿勢。正常

（4）Fetal posture

The posture of the fetus in the maternal body is

的胎勢為頭縱向、上位、胎兒前肢抱頭、後肢踢腹。

一般母畜分娩時，胎兒多是縱向、頭部前置，馬占 98%～99%，牛約占 95%，羊為 70%、豬為 54%。牛、羊懷雙胎時多為一個正生、一個倒生，豬往往是正倒交替產出。正常分娩的胎位、胎勢變化見圖 6-3。

called fetal posture, which is generally divided into extension and flexion. The normal fetal posture is head longitudinal, upper position, forelimbs embracing head and hind limbs kicking abdomen.

In general, the fetus is usually longitudinal with head presentation when in labor (about 98%-99% in horse, 95% in cow, 70% in sheep and 54% in sow). When cattle and sheep give birth to twins, most of the time one of which is in normal position and the other is inverted. Piglets are often delivered in alternate pattern of one in normal position and another in inverted position. Fetal position and fetal posture of normal parturition are shown in Figure 6-3.

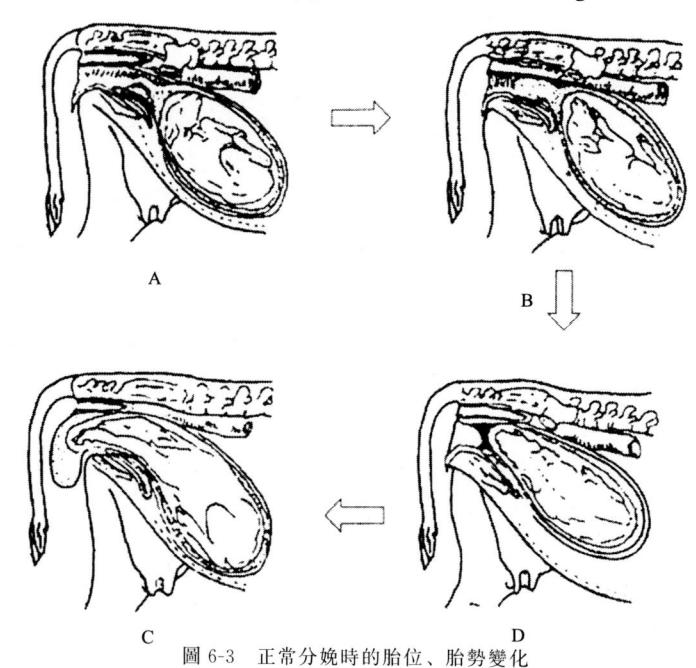

圖 6-3 正常分娩時的胎位、胎勢變化
Figure 6-3 Fetal position and fetal posture of normal parturition
A. 縱向下位　B. 頭、前肢後伸　C. 縱向側位　D. 縱向上位
A. longitudinal and inferior position　B. head and forelimbs rearward extension
C. longitudinal and lateral position　D. longitudinal and upper position

（二）分娩過程

分娩過程是從子宮肌和腹肌出現陣縮開始，至胎兒及其附屬物排出為止。分娩是有機連繫的完整過程，按照產道

3.2 Parturition process

Parturition starts with the myometrial and abdominal muscular contractions, until the fetus and its appendages are expelled. Parturition is a complete process,

暫時性的形態變化和子宮內容物的排出情況，習慣上將分娩過程分為子宮頸開口期、胎兒產出期和胎衣排出期 3 個階段。

1. 子宮頸開口期

從子宮出現陣縮開始，至子宮頸口完全開張為止。這一期只有陣縮而無努責。初產母畜表現不安、時起時臥、徘徊運動、頻頻舉尾。但經產母畜一般比較安靜。

2. 胎兒產出期

從子宮頸口完全開張至排出胎兒為止，陣縮和努責共同作用，而努責是排出胎兒的主要力量。當羊膜隨著胎兒進入骨盆入口，便引起膈肌和腹肌反射性和隨意性收縮。胎兒最寬部分的排出時間最長，特別是頭部通過骨盆及其出口時，母畜努責最強烈。

牛的產出期約為 6h，有的長達 12h；馬、驢產出期約為 12h，有的長達 24h；豬為 3～4h；綿羊為 4～5h；山羊為 6～7h。

3. 胎衣排出期

從胎兒排出後到胎衣完全排出為止。胎兒排出後，一方面母體胎盤的血液循環減弱，子宮黏膜腺窩的緊張性降低；另一方面胎兒胎盤的血液循環停止，絨毛膜上的絨毛體積縮小，間隙增大，使絨毛很容易從腺窩中脫落。

胎衣排出得快慢，因各種家畜的胎盤組織結構不同而有差異。瀰散型胎盤組織結合比較疏鬆，胎衣容易脫落，所以排出最快，豬、馬和驢的胎衣排出就

which can be divided into three stages: dilation of the cervix, expulsion of the fetus and expulsion of the placenta, according to the temporary morphological changes of the birth canal and the expulsion of uterine contents.

3.2.1 Dilation of the cervix

This stage begins with contraction of the uterus and ends with fully opened cervical orifice, only contracture without straining exists in this stage. The dam of first parturition is restless, having frequent sitting-ups and lying-downs, raising tails frequently. But the multiparous dams are usually quiet.

3.2.2 Expulsion of the fetus

This stage begins with fully opened cervical orifice and ends with the expulsion of the fetus, by the combined effect of contracture and straining, but straining is the main force to expel the fetus. When the fetus enclosed in the amnion enters the pelvic entrance, it causes contraction of the diaphragm and abdominal muscles. The widest part of the fetus takes the longest time to be expelled, as well as the strongest straining, especially when the head of the fetus passes through the pelvis and its outlet.

The duration of expulsion in cow, horse, donkey, sow, ewe and goat is respectively about 6-12, 12-24, 3-4, 4-5 and 6-7 hours.

3.2.3 Expulsion of the placenta

This stage begins with the expulsion of the fetus and ends with complete expulsion of placenta. After the fetus is expelled, the blood circulation of the maternal placenta and the tension of the uterine mucosal glandular fossa decrease, on the other hand, the blood circulation of the fetal placenta stops, and the volume of villi on the chorion shrinks and the gap increases, which makes the villi easily fall off from the glandular fossa.

The rate of placental expulsion varies with the structure of placenta in different livestock. The placentas of swine, horse and donkey belong to diffused-type placenta that the tissue structure is loosened and the placenta is easy to fall off. The expulsion period of

屬於這種情況，豬的胎衣排出期為10～60 min，馬和驢為5～90 min。牛、羊的胎盤屬於子葉型，結合比較緊密，子宮肌收縮時不容易影響腺窩，只有當母體胎盤組織的張力減小時，胎兒胎盤的絨毛才能脫落下來，所以需要時間較長。牛的胎衣排出期為2～8 h，綿羊為0.5～4 h，山羊為0.5～2 h。

四、難產

（一）難產分類

在分娩過程中，如果母畜產程過長或胎兒排不出體外，稱為難產。由於發生的原因不同，可將難產分為產力性難產、產道性難產和胎兒性難產3種。

（1）產力性難產。子宮陣縮及努責微弱、陣縮及破水過早和子宮疝氣等引起的產力不足而導致的難產。

（2）產道性難產。子宮捻轉，子宮頸、陰道及骨盆狹窄、產道腫瘤等引起的難產。

（3）胎兒性難產。胎兒過大，胎位、胎向和胎勢不正等引起的難產。

在上述三種難產中，以胎兒性難產最為多見，在牛的難產中約占75%，在馬、驢的難產中可達80%。

（二）難產的預防

（1）切忌母畜配種過早，若母畜尚未發育成熟，分娩時容易因骨盆狹窄造成難產。

（2）妊娠期間，要合理飼養母畜，給予營養全面的飼料，以保證胎兒發育

placenta is 10-60 minutes in swine and 5-90 minutes in horse and donkey. The placentas of cattle and sheep belong to cotyledon type that the structure is compact. The myometrial contractions of this placental type does not tend to affect the glandular fossa, that the villi of the placenta can fall off only when the tension of the maternal placenta tissue is reduced, so it takes a long time to expel the placenta. The expulsion periods of placenta in cattle, sheep and goats are 2-8 hours, 0.5-4 hours and 0.5-2 hours, respectively.

4 Dystocia

4.1 Classification of dystocia

Dystocia can be defined as the inability of the dam to expel neonates through the birth canal from the uterus, or prolonged labor that beyond the normal parturition time. According to the different causes, dystocia can be divided into three types: dystocia of labor force, dystocia of birth canal and fetal dystocia.

(1) Dystocia of labor force is caused by a deficiency of expulsive forces, such as weakness of contracture and straining, uterine hernia, ect.

(2) Dystocia of birth canal is caused by uterine torsion, birth canal tumor and stenosis of the cervix, vagina and pelvis.

(3) Fetal dystocia is caused by excessive fetal size and erroneous of fetal orientation, position and posture.

Among the three types of dystocia mentioned above, fetal dystocia is the most common, accounting for about 75% in cases of dystocia in cow and 80% in horse and donkey.

4.2 Prevention of dystocia

(1) The dam should not be prematurely bred. If the dam has not yet developed maturely, it is susceptible to dystocia due to pelvic stenosis during parturition.

(2) During gestation, the dam should be fed reasonably and given perfect nutrition to ensure the development of the fetus and the health of the dam and re-

專案六 接產與助產

Project VI Delivery and Midwifery

和母畜健康，減少難產發生的可能性。妊娠末期，要適當減少蛋白質飼料，以免胎兒過大。

（3）安排適當的使役和運動，可提高母畜對營養物質的利用，使全身及子宮肌的緊張性提高，有利於分娩時胎兒的轉位，防止胎衣不下及子宮復位不全等。

（4）做好臨產檢查，對分娩正常與否做出早期診斷。

duce the possibility of dystocia. At the end of pregnancy, reduce protein feed to avoid excessive fetal growth.

(3) By taking on proper tasks and exercises, the utilization of nutrients can be improved, and the tension of the whole body and uterine muscle can be improved, which is beneficial to the translocation of the fetus during delivery, and can prevent retention of the placenta and incomplete uterine reduction etc.

(4) Early diagnosis of whether the dam can undertake normal labour is made by prenatal examination.

任務 1 牛的接產與助產
Task 1 Delivery and Midwifery of Cattle

任務描述

王華大學畢業後到一家大型乳牛場工作，新員工職前培訓結束後，王華被分配到產房工作。他如何勝任產房的工作呢？

Task Description

After graduation, Wang Hua worked in a large dairy farm. He was assigned to the parturition room after the pre-job training. How could he be qualified for the parturition room?

任務實施

一、產前的準備

1. 產房

產房的基本要求為寬敞明亮、清潔乾燥、通風透光、沒有賊風、保溫良好和環境安靜。使用前應進行全面的清掃消毒，把產房的牆壁和地面、運動場、飼槽、分娩欄等打掃乾淨。

2. 器械及物品

準備好接產用具和藥品，如水盆、肥皂、紗布、藥棉、剪刀、助產繩、催產藥和消毒藥（如碘酒、酒精）、有

Task Implementation

1 Prenatal preparation

1.1 Parturition room

The basic requirements of parturition room are being spacious and bright, clean and dry, well-ventilated and transparent, no wayward wind blowing in, good heat preservation and quiet. The walls and floor of parturition room should be thoroughly cleaned and disinfected, the playground, as well as the trough and delivery stable before being used.

1.2 Instruments and articles

Prepare the delivery utensils and medicines, such as water basin, soap, gauze, medicinal cotton, scissors, midwifery rope, oxytocic and disinfectant (such as iodine

條件的牛場最好準備一套產科器械。

3. 助產人員

助產人員應具有一定的助產經驗，隨時觀察和檢查母牛的健康狀況，嚴格遵守接產操作程序。另外，由於母牛多在夜間分娩，所以要做好夜間值班。

4. 母牛

預產期前1～2週將待產母牛轉入產房。對臨產前的母牛，用溫水清洗外陰部，並用煤酚皂液或高錳酸鉀溶液徹底消毒。分娩時，讓母牛左側臥或站立，以免胎兒受瘤胃壓迫產出困難。

二、助產方法

母牛正常分娩時，一般無須人為干預，接產人員的主要任務是監視分娩狀況，並護理好新生犢牛，清除犢牛鼻腔、口腔內黏液，剪斷臍帶，擦乾皮膚，及時餵初乳。只有確定母牛發生難產時再進行助產。

（1）在胎兒進入產出期時，應及時確定胎向、胎位、胎勢是否正常。正常分娩的姿勢見圖6-4：一種是正生上位，頭頸和兩前肢伸直，且頭頸在兩前肢的上面；另一種是倒生上位，兩後肢伸直，胎兒以楔狀進入產道。如果胎向、胎位及胎勢均正常，不必急於將胎兒拉出，待其自然娩出。如果胎勢異常，可將胎兒推回子宮進行整復矯正。出現倒生時，應迅速拉出胎兒，免得胎兒腹部進入產道後，臍帶可能被壓在骨盆底下，造成窒息死亡。

tincture, alcohol). It is best to prepare a set of obstetric equipment for cattle farms with bigger budgets.

1.3 Midwife

Midwives should be experienced, strictly follow the procedures of delivery and be able to observe and check the health status of the cow at any time. In addition, the midwife should carry out night duties accordingly, as the cow usually give birth at night.

1.4 Cow

Transfer the predelivery cows to the parturition room 1-2 weeks before the expected parturition date, wash the vulva of the cow thoroughly with warm water and disinfect with lysol or potassium permanganate. When labouring, keep it in standing or left-lateral position to avoid fetal dystocia due to the pressure of rumen.

2 Midwifery methods

Human intervention is not needed when the cow labours normally. The main task of the midwife is to monitor the delivery situation, nurse the newborn calves, wipe the mucus from nasal cavity and mouth, cut the umbilical cord, dry the skin and feed colostrum in time. The midwifery is needed only when dystocia happens.

（1）When the fetus enters the delivery period, it should be determined whether the fetal orientation, position and posture are normal or not. The normal posture of the fetus during parturition is shown in Figure 6-4. One is head presentation, with the head and neck on the two forelimbs and stretch straightly, the other is inverted presentation, with the two hind limbs straight and the fetus entering the birth canal in a wedge-shape. If the fetal orientation, position and posture are normal, there is no need to pull the fetus out, just waiting for its natural delivery, but if those are abnormal, the fetus can be pushed back to the uterus for correction. When the fetus is in inverted presentation, the midwife should pulled out the fetus quickly to prevent the asphyxia and death, because the umbilical cord may be pressed under the pelvis after the abdomen enters the birth canal.

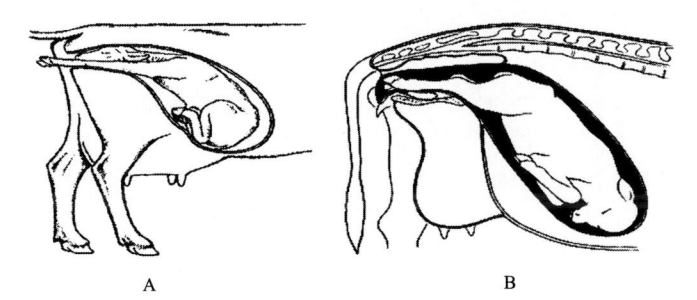

圖 6-4 正常分娩姿勢
Figure 6-4 Normal posture of the fetus during parturition
A. 正生上位　B. 倒生上位
A. head presentation and upper position　B. foot presentation and upper position

（2）當胎兒頭部已露出陰門外，如果排出時間過長，也應助產。如果胎膜尚未破裂（圖6-5），應及時撕破。拉出胎兒時，要注意保護好母牛的會陰部，以免發生陰道破裂或子宮脫出。如果破水過早，產道乾澀，可注入液狀石蠟進行潤滑。胎兒頭部尚未露出陰門外時，不要過早扯破羊膜，以防胎水流失，引起產道乾澀。

（2）Midwifery also should be implemented when the fetal head has been exposed outside the vulva for a long time. If the fetal membrane is not ruptured (Figure 6-5), midwives should tear it in time. When pulling out the fetus, attention should be paid to protect the perineum of the cow to avoid vaginal rupture or uterine prolapse. In order to prevent the dry and astringent of birth canal caused by the early loss of fetal water, liquid paraffin could be injected as lubricant, also midwives should not tear the amniotic membrane prematurely if the fetal head has not yet protruded out of the vulva.

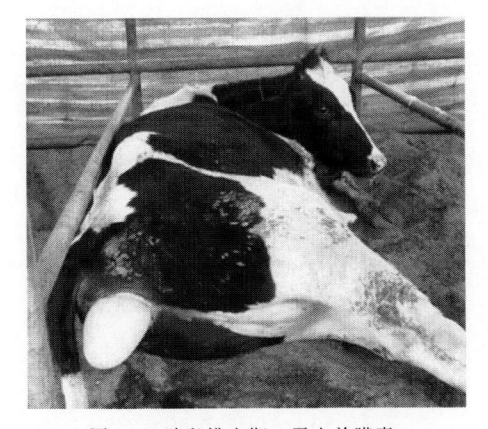

圖 6-5 胎兒排出期（露出羊膜囊）
Figure 6-5 Expulsion of the fetus (exposure of amniotic sac)

（3）當胎兒腹部通過陰門時，將手伸至其腹下握住臍帶根部和胎兒一起拉出，以免臍血管斷在臍孔內。

（4）如果母牛體弱，陣縮、努責無力，可用助產繩繫住胎兒兩前肢，趁母牛努責時順勢緩慢拉出胎兒。

（5）當母牛站立分娩時，應雙手接住胎兒，以免摔傷。

（6）胎衣排出後，及時檢查是否完整。如果胎衣排出不完整，說明母體子宮有殘留胎衣，要及時處理使胎衣完全排出。胎衣排出後，應立即取走，以免母牛吞食後引起消化紊亂。

三、難產救助方法

（一）難產的檢查

1. 產道檢查

該項目主要檢查產道是否乾燥、有無損傷、水腫或狹窄，子宮頸開張程度、有無損傷或瘢痕，骨盆腔是否狹窄及有無畸形、腫瘤等，並要注意產道內液體的顏色及氣味。

2. 胎兒檢查

常見的胎兒性難產有頭頸側彎、前肢屈曲、後肢屈曲、胎兒過大、胎兒畸形等（圖6-6）。另外，還要檢查胎兒的死活。正生時，助產人員可將手伸入胎兒口腔，輕拉舌頭或輕壓眼部或牽扯刺激前肢，注意觀察有無生理反應，如口吮吸、舌收縮、眼轉動、肢伸縮等；也可觸診頜下動脈或心區有無搏動。

（3）When the abdomen of the fetus goes through the vulva, the midwife should hold the root of the umbilical cord and pull it out, as well as the fetus, to prevent the umbilical vessel break in the umbilicus.

（4）If the cow is weak and the contracture and straining are powerless, the midwifery rope can be used to tie the two forelimbs of the fetus, then the fetus will be pulled out slowly while the cow is straining.

（5）When the cows are standing to labour, midwives should catch the fetus with both hands in order to avoid falling.

（6）Check the completeness of the placenta after it is discharged. If the placenta is incomplete, it indicates that there is residual placenta in the uterus of dams. After the placenta is discharged, it should be removed immediately to avoid being swallowed by the cow, which can induce digestive disturbance.

3 Rescue methods for dystocia

3.1 Examination of dystocia

3.1.1 Examination of birth canal

During this examination, the birth canal, cervix and the pelvis should be checked, for example, whether the birth canal is injured, dry, edematous or narrow, whether the cervix is fully dilated, injured or scarred, and whether the pelvic cavity is narrow and abnormal. In the meantime, pay attention to the color and odor of the liquid in the birth canal.

3.1.2 Examination of fetal

Fetal dystocia includes head and neck lateral bending, flexion of forelimb and hind limb, oversize and malformation of fetus (Figure 6-6). In addition, whether the fetus is dead should be examined. When the fetus is head presentation, midwives can palpate the submandibular artery or cardiac region for pulsation and extend their hands into the oral cavity of the fetus, gently pull fetal tongues or press fetal eyes or stimulate the forelimbs and pay attention to the physiological reactions, such as mouth sucking, tongue contraction, eye rotation,

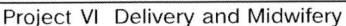

倒生時，觸診臍帶有無動脈搏動，也可牽拉刺激後肢注意有無反射活動，或將食指輕輕伸入肛門檢查其收縮反射。

limb contraction, etc. When the fetus is foot presentation, midwives can palpate umbilical cord for arterial pulsation and pay attention to the reflex activities of the fetus by pulling and stimulating hind limb or penetrating the anus with index finger.

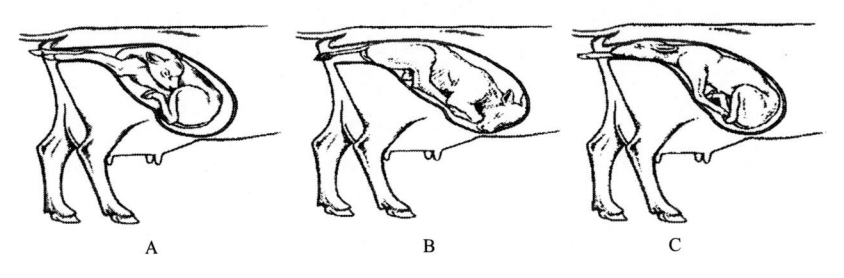

圖 6-6 常見胎兒性難產類型
Figure 6-6 Common types of fetal dystocia
A. 頭頸左彎 B. 前肢屈曲 C. 髖關節屈曲
A. left bending of head and neck B. forelimb flexion C. hip joint flexion

（二）難產的救助原則

當發生難產時，應鎮靜對待，查明原因，及時採取適宜措施。

（1）救助目的要明確，儘可能保證母仔雙方平安。在沒有可能的情況下，一般優先考慮母畜安全，但在胚胎移植時，則首先保證良種犢牛的存活。

（2）助產時，盡量避免產道感染和損傷，注意器械的嚴格消毒和規範使用。

（3）母畜橫臥分娩時，盡量使胎兒的異常部分向上，便於操作。

（4）為了便於推回或拉出胎兒，尤其是產道乾澀時，應向產道內灌注肥皂水或液狀石蠟潤滑劑。

（5）矯正胎兒反常姿勢時，盡量將胎兒推回子宮內進行矯正，因為產道空間有限，不易於操作。要掌握好推回胎

3.2 Rescue principles of dystocia

When dystocia occurs, midwives should treat it calmly, find out the cause and take appropriate measures in time.

(1) The purpose of the rescue should be clear and ensure the safety of both the dam and the young as far as possible. In general, the safety of the dam is the priority, but the survival of well-bred calves is the first priority in case of embryo transfer.

(2) During midwifery, midwives should try to avoid infection and injury of the birth canal and pay attention to strict disinfection and standardized use of instruments.

(3) Try to make the abnormal part of the fetus upward so as to easy operation when the dam lay in labor.

(4) In order to push back or pull out the fetus expediently, midwives should inject soapy water or paraffin oil into the birth canal especially when the birth canal is dry and astringent.

(5) When correcting abnormal fetal posture, midwives should try to push the fetus back into the uterus, as the limited space of the birth canal is not easy to op-

兒的時機，盡量選擇在陣縮的間歇期進行。矯正困難時，可進行剖腹術或截胎術。

（6）拉出胎兒時，應隨著母牛的努責而用力，並注意保護好母牛的會陰部，尤其是初產母牛，胎頭通過陰門時，會陰易發生撕裂。

（三）難產的救助方法

當發現難產時，要及時、果斷地進行救助。母牛難產時，先注入潤滑劑或肥皂水，再將胎兒順勢推回子宮，胎位矯正後，再順其努責輕輕拉出，嚴防粗暴硬拉。對於陣縮、努責微弱或子宮頸狹窄的母牛，可注射雌激素和催產素，刺激子宮收縮和宮頸開張。對於胎兒過大或難以矯正、無法助產的母牛，要及時進行剖腹產。

四、 新生犢牛護理

1. 清除黏液，注意保溫

犢牛剛出生後，立即清除其口腔、鼻腔周圍的黏液，盡快擦乾犢牛身體上的黏液，以防受涼。如犢牛呼吸困難或停止，迅速抓住犢牛後肢將其倒吊，輕拍犢牛胸部和背部，同時注意清除口鼻倒流出的黏液，直到呼吸順暢。

2. 斷臍消毒

臍帶未斷裂時，捏住臍帶基部，用消毒剪刀距腹部 6～8cm 處剪斷。斷端用 5％ 碘酒消毒，一般不結紮，以利於乾燥癒合。

erate. The intermittent period of contracture is a good time to push back the fetus. When it is difficult to correct the fetal posture, cesarean section or fetotomy can be performed.

(6) Midwives should pull out the fetus with the straining of the cow and pay attention to protect the perineum of the cow, especially when the fetal head of the primiparous cow passes through the vulva, the perineum is easy to tear.

3.3 Rescue methods for dystocia

When dystocia occures, midwives first injects lubricant or soapy water into uterine lumen and around the fetus, then push the fetus back into the uterus to correct the fetal position, finally gently pull the fetus out with the straining of cow. For cows with weak constrictio, nutria, or narrow cervix, estrogen and oxytocin can be injected to stimulate uterine contraction and cervix opening. For the cows whose fetuses are too large or difficult to correct and unable to accept midwifery, cesarean section should be carried out in time.

4 Nursing of newborn calves

4.1 Clean up mucus and keep warm

Immediately wipe mucus from the newborn's nostrils and mouth with a clean, dry cloth, and gently rub its body to prevent it from catching cold after birth. If the calf has dyspnea or stops breathing, hang the calf by its hind legs immediatly, pat the chest and back of the calf, and pay attention to clean up mucus from the mouth and nose until breathing is smooth.

4.2 Rupture umbilical cord and disinfection

When the umbilical cord is not broken, midwife should pinch its root and cut the umbilical cord 6-8 cm from the calf's body, then dip the umbilical cord in 5% tincture of iodine to prevent infection, which is usually not ligated to facilitate drying and healing.

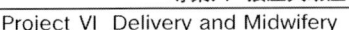

3. 儘早餵初乳

飼餵初乳的時間越早越好，一般在產後 0.5 h 內，最長不能超過 1 h。初乳飼餵量為犢牛體重的 8%～10%，大型牧場為便於操作一般統一灌服 4 L。

4. 做好記錄工作

記錄是牛場管理的重點，從新生犢牛開始就要做好相關記錄，如犢牛耳號、出生日期、體重、母牛的胎次及產犢難易程度等相關資訊。

五、產後母牛的護理

1. 注意胎衣和惡露排出情況

要觀察胎衣排出情況，母牛分娩後 8 h 內，胎衣一般可自行排出，若超過 24 h 仍不排出，則稱為胎衣不下。根據胎衣滯留時間的長短，可採取藥物治療和手術剝離兩種方法。注意惡露的排出情況，惡露一般會持續 10～14 d。惡露排完，說明子宮已復原。如果母牛在產後 3 週仍有惡露排出或惡露腥臭，表示有子宮感染，應及時治療。

2. 注意產後母牛外陰部和陰道的清潔消毒

對產後 3～5 d 的母牛，應每天用溫水、肥皂水、1%～2% 來蘇兒或 0.1% 高錳酸鉀溶液清洗母牛外陰部並擦乾。

3. 加強飼養管理

母牛分娩之後，要及時供給足夠的水和麩皮湯或益母草紅糖水等，有利於胎衣的排出和子宮的復原。產後 1～2 d 天的母牛還應繼續飲用溫水，飼餵品質好、易消化的飼料，投料不宜過多，一般 5～6 d 後可以逐漸恢復正常飼養。

4.3 Feed colostrum as early as possible

The calf should be fed colostrum within half an hour after birth, and no more than 1 hour at the latest. Depending on the weight of the calf, you need to feed 8%-10% percent of the calf weight. In large-scale ranch, usually uniformly fed 4L colostrum generally.

4.4 Complete recording task

Recording is the key point of cattle farm management. It is best to record relevant data, such as the ear mark, birth date, weight, gravidity of cow and difficulty levels in calving when they are first born.

5 Nursing of postpartum cow

5.1 Discharge of placenta and lochia

The cow should discharge the placenta within 8 hours after parturition. If the cow has not discharged the placenta within 24 hours after calving, it is called the remained placenta. Medication and surgical stripping can be used according to the length of the stay time of the placenta. The lochia usually lasts for 10-14 days. The exhaustion of lochia indicates that the uterus is restored. If the cow still discharge lochia 3 weeks after parturition or lochia is stinking, it indicates that there is uterine infection and should be treated in time.

5.2 Cleaning and disinfection of vulva and vagina in postpartum cows

The vulva of the cow should be cleaned with warm water, soapy water, 1%-2% lysol or 0.1% potassium permanganate and dried every day within 3-5 days after parturition.

5.3 Strengthen feeding and management

It is necessary to supply sufficient water and bran soup or leonurus brown sugar water in time to facilitate the discharge of placenta and recovery of the uterus in cows after parturition. Cows should continue to drink lukewarm water and to be fed good quality and digestible feed within 1-2 days after parturition. Do not feed too much. Generally, the cows can gradually return to normal feeding 5-6 days after parturition.

任務 2　羊的接產與助產
Task 2　Delivery and Midwifery of Sheep

任務描述

做好母羊的接產與助產工作，對於維護母羊健康、提高羔羊成活率均具有重要意義。母羊分娩前有哪些預兆？如何做好新生羔羊的護理工作？

Task Description

It is of great significance to do well in delivery and midwifery of ewes for maintaining the health of ewes and improving survival rate of lamb. What are the signs before parturition? How to nurse newborn lambs?

任務實施

一、產前的準備

1. 產房

產前 3～5d 將產房、運動場、飼槽、分娩欄等清掃乾淨。為了讓分娩母羊熟悉產房環境，在臨產前 2～3d 就應將其圈入產房，確定專人管理，隨時觀察。

2. 接產人員

接產人員應具有豐富的接羔經驗，熟悉母羊的分娩規律，嚴格遵守操作規程。

3. 用具及器械

肥皂、毛巾、藥棉、紗布、聽診器、細繩、剪刀、鑷子、常用產科器械及必需藥品（如酒精、碘酒、新潔爾滅、縮宮素、抗生素等）。

二、助產方法

（1）母羊臨產前，首先剪去乳房周圍和後肢內側的毛，用溫水洗淨乳房，並擠出幾滴初乳，再將母羊的尾根、外陰部和肛門洗淨，然後用 1% 來蘇兒消毒。

Task Implementation

1　Prenatal preparation

1.1　Parturition room

Parturition room should be cleaned and disinfected 3-5 days before parturition and the ewes should be transferred to the parturition room to familiarize themselves with the environment 2-3 days before parturition. It should be managed by particular person, who can observe and monitor at any time.

1.2　Midwife

Midwives should be experienced, familiar with the rule of ewe delivery, and can strictly follow the operating procedures.

1.3　Instruments and supplies

Soap, towel, cotton, gauze, stethoscope, string, scissors, tweezers, commonly-used obstetric instruments and essential drugs (such as alcohol, iodine, neogeramine, oxytocin, antibiotics, etc.).

2　Midwifery methods

(1) Before parturition of ewe, midwives should first shear the wool around the breast and inside the hind limbs, then wash the breasts with warm water and squeeze out a few drops of colostrum, finally wash the tail root, vulva and anus and sterilize them with 1% lysol.

（2）母羊正常分娩時，一般不予干擾，最好讓其自行分娩，一般在胎膜破裂、羊水流出後幾分鐘至 30min，羔羊即可產出。正常分娩時，羔羊兩前肢夾頭先產出，其餘隨後產下（圖 6-7）。

(2) Ewes usually deliver lamb within a few minutes to 30 minutes after rupture of membranes and outflow of amniotic fluid. In normal labour, two forelimbs and head come out first and then followed by the rest(Figure 6-7).

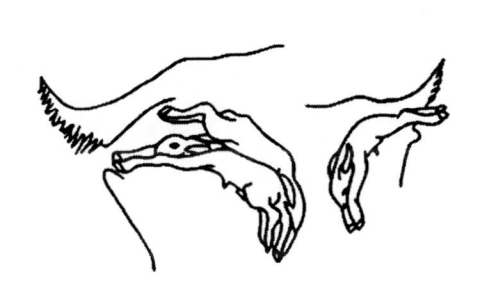

圖 6-7　羊的正常胎位
Figure 6-7　normal fetal position of sheep

（3）產雙羔時，先後間隔 5～30min，但也偶有長達數小時以上的。因此，當母羊產出第一羔後，如仍表現不安、臥地不起，或起立後又重新躺下、努責等，可用手掌在母羊腹部前方適當用力向上推舉，如果可觸摸、感覺到光滑的羔體，說明分娩還未結束。

（4）當母羊產道較為狹窄或體乏無力時，需要人工助產。其操作方法：助產人員在母羊體軀後側，用膝蓋輕壓其肷部，等羔羊前端露出後，用手推動母羊會陰部，待羔羊頭部露出後，然後一手托住頭部一手握住前肢，隨母羊努責向後下方拉出胎兒。

(3) Double lambing is produced at intervals of 5-30 minutes, but sometimes more than a few hours. Therefore, after the ewe produces the first lamb, if the ewe is still restless, lying down, or standing up and lying down alternatively, and straining, etc., midwives can properly put their palms on the abdomen of the ewes and then push them upward. If a smooth lamb body can be touched, it means the parturition is not over yet.

(4) Artificial midwifery is needed when the birth canal of the ewe is too narrow or the ewe is weak.Operation methods: the midwife first stands on the back of ewe's body and gently presses the buttock of the ewe with knees, then pushes the perineum when the front of the lamb is exposed, next holds the head with one hand and holds the forelimbs with the other hand when the head of lamb is exposed, finally pull the fetus out in sync with the straining of the ewe.

三、難產救助方法

3　Rescue methods for dystocia

（1）胎兒過大、陰道狹窄、羊水過早流失等原因形成的難產，先用凡士林

(1)Dystocia can be caused by oversized fetus, vaginal stenosis and premature loss of amniotic flu-

或液狀石蠟潤滑陰道，然後把胎兒的兩前肢拉出來再送進產道去，反覆三四次擴大陰門後，配合母羊陣縮補加外力牽引，幫助胎兒產出。

（2）胎位、胎向、胎勢不正時（圖6-8），接羔人員應在母羊陣縮間歇時，用手將胎兒輕輕推回腹腔，手也隨著伸進陰道，用中指、食指矯正異常的胎位、胎向，並協助將胎兒拉出。

（3）因子宮陣縮及努責微弱引起的產力性難產，可肌內注射或靜脈注射適量催產素。

（4）因陰道、陰門狹窄和子宮腫瘤等引起的產道性難產，可在陰門兩側上方將陰唇剪開1～2cm，將陰門翻起同時壓迫尾根基部，以使胎頭產出而解除難產。如果母羊的子宮頸過於狹窄或不能擴張，應施行剖腹產手術。

（5）雙羔同時楔入產道時（圖6-9），助產人員應將消毒後的手臂伸入產道將一個胎兒推回子宮內，把另一個胎兒拉出後，再拉出推回的胎兒。如果雙羔各將一個肢體伸入產道，形成交叉的情況，則應先辨明關係，順手觸摸肢體與軀幹的連接，分清肢體的所屬，將其中一個胎兒推回子宮內，把另一個胎兒胎向、胎勢調整好後拉出，再拉出推回的胎兒。

四、 新生羔羊護理

1. 清除黏液

羔羊出生後，用手先將其口腔、鼻

id. Midwives should lubricate the vagina with vaseline or paraffin oil first, then pull out the fetal forelimbs and send them back to the birth canal. After repeating three or four times, the vulva is expanded, and the fetus will be pulled out with the contracture of the ewe and external traction.

（2）Dystocia can be caused by abnormal fetal orientation, position and posture (Figure 6-8). The midwife should make use of the contracture intervals of ewes to gently push the fetus back to the abdominal cavity with hands and correct the abnormal fetal position with middle finger and index finger, then pull the fetus out.

（3）Labour force dystocia caused by weakness of uterine contracture and straining can be rescued by injecting oxytocin intramuscularly or intravenously.

（4）Birth canal dystocia, which caused by stenosis of vagina and vulva or uterine tumors, can be relieved by cutting the labia 1-2 cm above both sides of the vulva, turning the vulva upside down and pressing the base of the tail root at the same time, so as to deliver the fetal head and resolve dystocia. If the cervix of the ewe is too narrow or unable to dilate, the cesarean section should be performed.

（5）When the twin lambs enter the birth canal at the same time (Figure 6-9), the midwife should extend the sterilized arm into the birth canal, push one fetus back into the uterus, pull one fetus out and then pull the other fetus out. If the twin lambs each extends a limb into the birth canal to form a crossover position, the midwife should first touch the connection between the limb and the trunk, distinguish the ownership of the limb. Then push one fetus back into the uterus, adjust the fetal position of the other correctly and pull it out, then the first one.

4 Nursing of newborn lambs

4.1 Clean up the mucus

After the lamb is born, the mucus in its mouth and

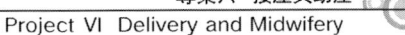

腔裡的黏液掏出擦淨，以免因呼吸困難、吞食羊水而引起窒息或異物性肺炎。其餘部位的黏液讓母羊舔乾，有利於母羊認羔。如天氣寒冷，要用乾淨布迅速將羔羊身體擦乾，以免受涼。

nasal cavity should be wiped out to avoid asphyxia or foreign body pneumonia caused by dyspnea or swallowing amniotic fluid. It is beneficial for ewes to recognize lambs that the ewes lick off the remaining mucus. If the weather is cold, midwives should dry the lamb quickly with a clean cloth, so as to not catch a cold.

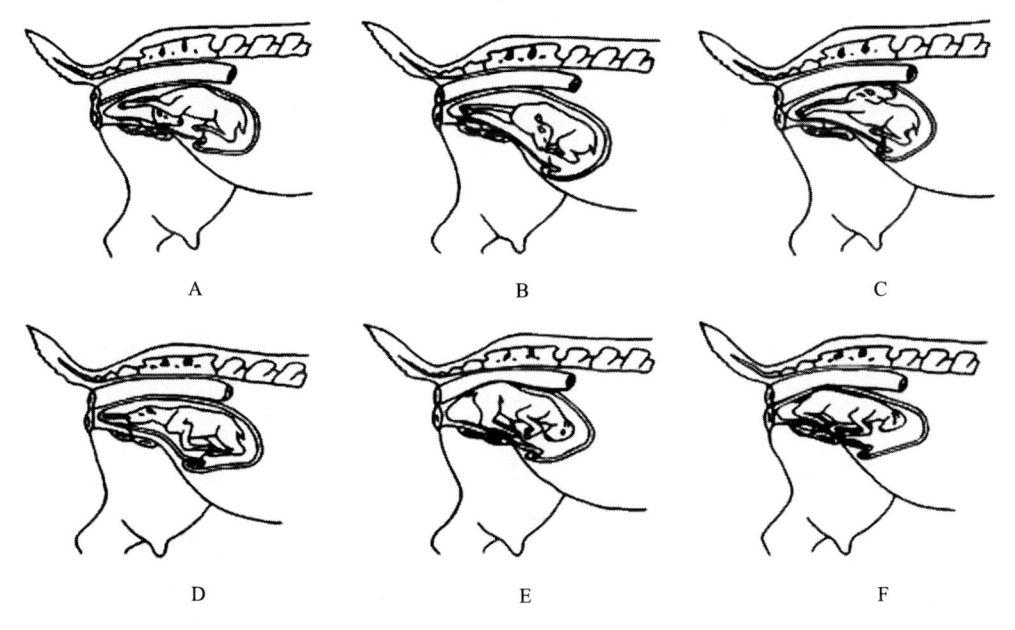

圖 6-8　羊部分胎勢不正

Figure 6-8　Abnormal fetal position in sheep

A. 頭向後仰　B. 頭向下彎　C. 頭頸側彎　D. 前肢彎曲　E. 坐骨前置　F. 跗部前置

A. head backward　B. head down bend　C. head and neck lateral bend
D. forelimb curvature　E. ischium presentation　F. tarsal presentation

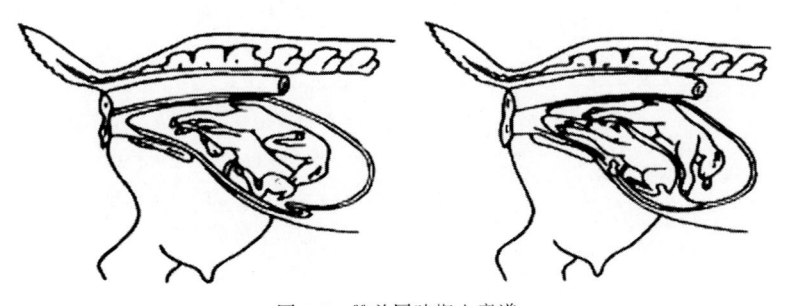

圖 6-9　雙羔同時楔入產道

Figure 6-3　Twin lambs wedge into the birth canal

2. 斷臍

羔羊出生後臍帶可自行斷裂，或在臍帶停止波動後距腹部 4～6cm 處用手擰斷，斷端用 3%～5% 碘酒消毒。

3. 盡快吃初乳

羔羊出生後，一般十幾分鐘即能站立，如果一胎多羔，不能讓第一個羔羊把初乳吃淨，應使每個羔羊都能吃到初乳。

4. 假死羔羊處理

有些羔羊出生後不呼吸或呼吸微弱，但心臟仍有跳動，這種現象稱為假死。假死羔羊的搶救方法：①先清除呼吸道內的黏液或羊水，然後用酒精棉球或微量碘酒滴入羔羊鼻孔以刺激羔羊呼吸，或向羔羊鼻孔吹氣、噴煙使之甦醒；②將羔羊兩後肢提起懸空並輕輕拍打其背胸部；③將假死羔羊放平，兩手有節律地推壓羔羊胸部兩側。

5. 做好記錄

將羔羊編號，育種羔羊秤量出生重，按要求填寫羔羊出生登記表。

五、 產後母羊的護理

注意觀察胎衣排出情況，羊的胎衣通常在分娩後 2～4h 內排出。產羔母羊要保暖防潮，產後 1h 左右，給母羊飲水，一般為 1～1.5 L，水溫 25～30℃，忌飲冷水，可加少許食鹽、紅糖和麥麩。剪去母羊乳房周圍的長毛，用溫毛巾擦洗乳房，並擠掉少量乳汁，幫助羔羊吃上初乳。

4.2 Rupture umbilical cord

The umbilical cord can rupture automatically, or it can be broken off by hand 4-6 cm away from the abdomen after the umbilical cord stops fluctuating, and the broken ends should be disinfected with 3%-5% iodine tincture.

4.3 Feed colostrum as early as possible

The lamb can stand up ten minutes after birth. If the ewe is polytocous, the first lamb should not be allowed to eat up the whole colostrum reserve, and make sure each lamb can eat colostrum fairly.

4.4 Treatment of the thanatoid lamb

Some lambs do not breathe or breathe weakly after parturition, but the hearts still beat. This phenomenon is called thanatosis. Rescue methods of thanatoid lambs: First, clean up mucus or amniotic fluid in the respiratory tract, then drip alcohol or trace iodine liquor into the nostrils to stimulate its breathing, or blow air into the nostrils to resuscitate. Hung the hind limbs of lambs upside down with one hand and gently pat its back and chest with the other hand. Lay down the thanatoid lamb flat on ground and press both sides of its chest rhythmically with both hands.

4.5 Complete recording

Midwives should number the lamb, weigh the birth weight of the breeding lamb, and fill in the birth registration form as required.

5 Nursing of postpartum ewes

The placenta of sheep is usually expelled within 2-4 hours after parturition. Ewes should drink water about one hour after parturition, the water volume is generally 1-1.5 litre and the temperature is 25-30℃, no drinking cold water, a little salt, brown sugar and wheat bran can be added to the water. The midwife should cut the long hair around the ewe's udder, scrub the udder with a warm towel, squeeze out a little milk, and help the lamb to eat colostrum.

任務3 豬的接產與助產
Task 3　Delivery and Midwifery of Pigs

任務描述

母豬分娩是養豬生產中最繁忙、最細緻、最易出問題的環節。為了保證母豬安全分娩和提高仔豬成活率，需熟練掌握接產與助產技術。對於假死仔豬，該如何搶救？

Task Description

Sow parturition is the busiest, most meticulous and most problematic link in pig production. In order to ensure safe delivery and improve the survival rate of piglets, it is necessary to master the techniques of delivery and midwifery. How to treat the thanatoid piglets?

任務實施

一、產前的準備

1. 產房

分娩前5～7d準備好產房，產房要求溫暖乾燥、清潔衛生、舒適安靜、陽光充足、空氣清新，並用2%～5%來蘇兒或2%～3%氫氧化鈉溶液噴霧消毒。產前一週將母豬轉移到產房中。

2. 接產用具

接產時應準備好下列用具：產仔哺育記錄卡、手術剪、毛巾、碘酊、高錳酸鉀、結紮線、耳號鉗、斷齒鉗等。

二、助產方法

仔豬產出後，立即用清潔毛巾擦去其口、鼻中的黏液，然後再擦乾全身。斷臍後，進行秤重、編號、剪齒、斷尾等，併作好記錄。

三、難產救助方法

母豬一般不發生難產，如果出現反

Task Implementation

1　Prenatal preparation

1.1　Parturition room

The preparation requirements of the parturition room is similar to those of cattle and sheep, which should be warm and dry, clean and sanitary, comfortable and quiet, sunny and fresh. The sows are transferred to the parturition room one week before parturition.

1.2　Delivery tools

The following instruments should be prepared during delivery: record cards for farrowing, surgical scissors, towels, iodine tincture, potassium permanganate, ligation line, forceps for marking ear and breaking teeth, etc.

2　Midwifery methods

After the piglet is born, wipe the mucus from the mouth, nose and whole body. A lot of work should be done immediately after cutting umbilical cord, such as weighing, numbering, cutting teeth and tail, recording ect..

3　Rescue methods for dystocia

Sows generally do not suffer from dystocia and

覆陣痛、努責、呼吸及心跳加快等症狀即為難產。發現母豬難產時，應馬上進行助產，常用「推、拉、注、掏、剖」五字助產技術。

1. 推

如果胎位不正，可採取「推」的辦法。即接產者用雙手托住母豬後腹部，隨母豬努責時向臀部方向用力推，但不可硬壓，調整好胎位後再行助產。

2. 拉

看見仔豬的頭或腿部時出時進時，只要胎位正，就可以拉出仔豬。

3. 注

母豬分娩過程中如出現產力不足，可肌內注射催產素 10～20 IU，20～30 min 起效。

4. 掏

產不出仔豬時，需要人工掏出。趁母豬努責間歇時，接產員將手指合攏成圓錐形伸入母豬產道，摸到仔豬的適當部位，將仔豬慢慢拉出。如果有兩頭仔豬同時擠在一起，先將其中一頭推回，再抓住另一頭隨母豬努責拉出，掏出一頭仔豬後，如轉為順產，就可不再繼續掏。掏完後，用手把 40 萬 IU 的青黴素抹入陰道內，以防陰道炎。

5. 剖

採取上述方法後仔豬還生不出來，就應進行剖腹產術。

四、新生仔豬護理

1. 注意觀察臍帶

臍帶斷端一般於仔豬出生後 1 週左右乾縮脫落。此期應注意觀察，勿

the symptoms of dystocia include repeated pains, straining, rapid breathing and heartbeat, etc. If it occurs, the midwifery should assist immediately by using some techniques, which can be summarized as 「push, pull, inject, dig and caesarean」.

3.1 Push

This method can be used if the fetal position is not correct. Midwives hold the postabdomen of the sow with both hands and push the piglets to the buttocks with the straining of sows to adjust the fetal position.

3.2 Pull

Piglets can be pulled out when the fetal position is right, especially when the head or leg of the fetus coming in and out repeatedly.

3.3 Inject

If the force of labour is insufficient during delivery, oxytocin can be injected intramuscularly for 10-20 IU and it will take effect in 20-30 minutes.

3.4 Dig

If the piglet can not be delievered, midwives need to take them out. Midwives close their fingers into a conical shape and extend into the birth canal during the contracture interval of sows, touch the appropriate parts of the piglet and slowly pull them out. If two piglets are squeezed together, midwives should send one back to the abdomen first, then grab the other one and pull it out with the straining of sows. If the sows turn to normal labour, midwives can stop digging. After digging out, 400 000 IU of penicillin should be wiped into the vagina by hand to prevent vaginitis.

3.5 Caesarean

If the piglet delivery is still impossible after using the aforementioned methods, cesarean section should be performed.

4 Nursing of newborn piglets

4.1 Pay attention to the umbilical cord

It takes about one week after parturition that the broken ends of umbilical cord shrink and abscise.

使仔豬間互相舔吮引起感染發炎，如臍血管或臍尿管關閉不全，應進行結紮處理。

2. 早吃初乳

新生仔豬儘早吃好、吃足初乳，最晚不得超過生後 2h。吃乳前母豬乳頭用 0.1%高錳酸鉀擦洗消毒，用手擠壓各乳頭棄去最初擠出的乳汁，然後給仔豬吮吸。

3. 保溫

初生仔豬皮薄毛稀，皮下脂肪少，體溫調節能力差，對極端溫度反應敏感。尤其是在冬季，應密切注意防寒保溫，確保產房溫度適宜。

4. 假死仔豬急救

個別仔豬產出後不呼吸，但心臟仍在跳動，手指輕壓臍帶根部可摸到脈搏。急救方法如下：

（1）先除去仔豬口腔、鼻腔內的黏液，然後兩手反覆伸屈仔豬的兩前肢和後肢，直到有呼吸為止。

（2）向假死仔豬鼻內或嘴內用力吹氣，促其呼吸。

（3）用左手提起仔豬兩後肢，頭向下，再用右手拍胸敲背，可救活仔豬。

（4）也可用酒精、碘酒、氨水等塗在仔豬鼻端，刺激其鼻腔黏膜恢復呼吸。

5. 仔豬寄養

為了充分利用生產母豬的所有有效乳頭，最大限度地保持吃乳仔豬群的均勻度，需要將一窩中超過母豬乳頭數的仔豬進行寄養。寄養前要吃足初乳，一

During this period, piglets should not be allowed to lick each other to prevent infection and inflammation. If the shrink of umbilical vessel and urachus is incomplete, ligation should be performed.

4.2 Feed colostrum as early as possible

The newborn piglets should eat enough colostrum no more than 2 hours after parturition. The nipples of the sow are scrubbed and disinfected with 0.1% potassium permanganate before feeding, and should be squeezed by hand to discard the original milk, before being sucked by piglets.

4.3 Keep warm

Newborn piglets have thin fur, less subcutaneous fat, poor body temperature regulation and are sensitive to extreme temperature. Especially in winter, midwives should pay more attention to ensure that the temperature of parturition room is appropriate.

4.4 Treatment of thanatoid piglets

Some piglets do not breathe after parturition, but their hearts are still beating. A pulse can be felt by pressing finger on the root of the umbilical cord gently. First aid methods are as follows:

(1) First, remove the mucus from the mouth and nasal cavity, then stretch and flex the forelimbs and hind limbs of the piglet repeatedly until it is breathing.

(2) Blow air into the nose or mouth of thanatoid piglets to stimulate them breathing.

(3) Hang the hind limbs of piglets upside down with one hand and gently pat its back and chest with the other hand.

(4) Smear alcohol, iodine tincture and ammonia water on the nasal tip of piglets to stimulate nasal mucosa to restore breathing.

4.5 Foster nursing of piglets

If the litter number of sows exceeds the number of nipples in sows, foster nursing is required. Midwives should feed enough colostrum and smear milk or urine of nurse's sows to the piglets before foster nursing. Foster nursing is generally performed about 2-3

般在出生後2～3d進行。寄養前，在寄養仔豬身上塗抹保姆母豬的乳汁或尿液。

6. 固定乳頭

為了減少仔豬在吸乳時的爭鬥，提高仔豬均勻度，促使母豬乳頭充分發育，提高泌乳量，要及時固定乳頭。一般母豬的第一對胸部乳頭泌乳最多，第2～6對乳頭的泌乳量依次遞減，因此，在出生後的第一次吸乳時就要初步固定乳頭，讓強仔靠後，弱仔靠前。

7. 剪齒

仔豬出生時就有上下各兩對銳利的犬齒，如不及時修剪，常因仔豬爭搶乳頭而把母豬的乳頭咬傷。同時，仔豬也因搶占乳頭相互咬傷而感染。

8. 定位訓練

新生仔豬在生後一至數小時，最晚第二天就會排泄糞尿。在預定排糞地點，放置少許母豬糞便，可誘導仔豬排糞定位。仔豬排糞尿多在吃乳前後，應在此時注意訓練定位排糞。

五、 產後母豬的護理

產後2～3 d內，餵料不宜過多，應用易消化的飼料調成粥狀飼餵。餵量逐漸增加，經5～7 d後可按哺乳母豬的標準飼餵。母豬分娩後及時清洗消毒其乳房和外陰部，保持產房的清潔衛生。同時應保持產房安靜，保證母豬有充分的休息時間。

days after parturition.

4.6 Fixed nipple

In order to reduce the struggle of piglets during sucking, and improve the evenness for each piglet to get fed properly, as well as promote the full development of nipples and increase the lactation volume, it is necessary to fix nipples in time. Generally, the milk yield of the first pair nipples of sows is the most, the milk yield of the second to sixth pairs of nipples decreases successively. Therefore, the nipples should be fixed initially during the first lactation at birth, that arrange the strong piglets behind and the weak piglets ahead.

4.7 Teeth clipping

Piglets are born with two pairs of sharp canine teeth. If the canine teeth are not clipped in time, the piglets often bite the nipples of the sow as they fight for the nipples. At the same time, piglets also bite each other because of fighting for nipples, thus leading to infection with sores.

4.8 Orientation training

Newborn piglets excrete feces and urine from several hours to the next day after birth. A little excrement of sows is placed at the designated place, that can induce the location of excrement of piglets. The best time for the location of defecation training is before and after the piglets sucking milk.

5 Nursing of postpartum sows

The digestible forage should be manufactured into mashy porridge-like food to feed sows within 2-3 days after parturition, and then the feeding quantity increases gradually.

After 5-7 days, the feed can be fed according to the standard of lactating sows. After delivery, breasts and genitals of the sows should be cleaned and disinfected, as well as keeping the delivery room clean. At the same time, the delivery room should be kept quiet, so that the sows can have a good rest.

專案七 繁殖力評價
Project Ⅶ Evaluation of Fecundity

⬥ 專案導學

繁殖力是評定種用畜禽生產力的主要指標，合理評價畜禽的繁殖力，能客觀瞭解畜禽的繁殖狀況，為制定繁殖管理措施提供依據。

⬥ Project Guidance

Fecundity is the main indicator to evaluate the productivity of breeding animals. Reasonable evaluation of fertility can objectively understand the reproductive status and provide a basis for formulating reproductive management measures.

⬥ 學習目標

>>> 知識目標

• 理解繁殖力評定指標的含義及統計方法。

• 熟悉各種畜禽的正常繁殖力。

>>> 技能目標

• 能根據養殖場的繁殖資料，統計分析各繁殖力指標，並進行綜合評價。

• 結合生產實際，擬定養殖場的繁殖管理措施。

⬥ Learning Objectives

>>> Knowledge Objectives

• To understand the meaning and statistical methods of fertility assessment indicators.

• Be familiar with the normal fecundity levels of various livestock and poultry.

>>> Skill Objectives

• According to the breeding data of the farm, analyze various reproductive indicators and evaluate comprehensively.

• Combining with the actual production, formulate the breeding management measures of the farm.

⬥ 相關知識

一、繁殖力

繁殖力是指畜禽維持正常生殖機能、繁衍後代的能力。對於種用畜禽來說，繁殖力即是生產力。公畜（禽）的繁殖力主要體現在性成熟早晚、性慾強

⬥ Relevant Knowledge

1 Fecundity

Fecundity refers to the ability of livestock and poultry to maintain normal reproductive function and reproduce offspring. For breeding animals, fecundity is productivity. Fecundity of male animals is mainly re-

弱、交配能力及精液品質等，母畜（禽）的繁殖力主要體現在性成熟早晚、發情排卵情況以及配種妊娠、分娩和哺乳等生殖活動。

二、繁殖力的評定指標

(一) 家畜繁殖力評定指標

1. 受配率

本年度內參加配種的母畜數占適繁母畜數（不包括因妊娠、哺乳及各種卵巢疾病等原因造成空懷的母畜）的百分率。

2. 受胎率

受胎率是指妊娠母畜數占參加配種母畜數的百分率。在受胎率統計中又分為總受胎率、情期受胎率、第一情期受胎率和不返情率。

總受胎率指本年度妊娠母畜數占本年度內參加配種母畜數的百分率。情期受胎率指妊娠母畜數占配種情期數的百分率。第一情期受胎率指第一情期配種的妊娠母畜數占第一情期配種母畜數的百分率。不返情率指配種後一定時間內（如 30d、60d、90d 等）未出現發情的母畜數占參加配種母畜數的百分率。

3. 分娩率

分娩率是指本年度內分娩母畜數占妊娠母畜數的百分率。其大小反映母畜妊娠品質的高低和保胎效果。

4. 產仔率

產仔率是指母畜的產仔（包括死胎）數占分娩母畜數的百分率。單胎動物如牛、馬、驢等，產仔率一般不會超過 100%，生產上多使用分娩率。多胎

flected in sexual maturation, sexual desire, mating ability and semen quality. Fecundity of female animals is mainly reflected in sexual maturation, estrus and ovulation, pregnancy, delivery and lactation.

2 Indicators of evaluating fecundity

2.1 Indicators for evaluating livestock fecundity

2.1.1 Breeding rate

Breeding rate refers to the number of bred females accounted for brood matron in one year (excluding unpregnant females that are caused by pregnancy, lactation and various ovarian diseases).

2.1.2 Conception rate

Conception rate refers to the number of pregnant females accounted for the bred females. In the statistics of conception rate, it can be divided into total conception rate, cycle conception rate, first cycle conception rate and non-return rate.

Total conception rate refers to the number of pregnant females accounted for bred females in one year. Cycle conception rate refers to the number of pregnant females accounted for total cycles. First-cycle conception rate refers to the number of pregnant females in the first cycle accounted for the females bred in the first cycle. Non-return rate refers to the number of females who do not estrus within a certain period of time after breeding (30d, 60d, 90d, etc.) accounted for bred females.

2.1.3 Delivery rate

It refers to the number of delivered females accounted for pregnant females in the current year. It reflects the quality of pregnancy and the effect of fetal protection.

2.1.4 Farrowing rate

It refers to the number of the newborns (including stillbirths) accounted for the delivered females. Single-born animals (such as cattle, horses, donkeys, etc.) generally do not exceed 100% of the litter size. Delivery rate is often used in production rather than farrowing

動物如豬、山羊、兔等，產仔率均會超過 100%，生產上多使用產仔率。

5. 仔畜成活率

仔畜成活率指本年度內斷奶成活的仔畜數占本年度產出活仔畜數的百分率。其大小反映仔畜的培育情況。

(二) 家禽繁殖力指標的統計

1. 產蛋量

產蛋量是指某群家禽一年內平均產蛋的枚數。

2. 受精率

受精率是指種蛋孵化後，經第一次照蛋確定的受精蛋數占入孵蛋數的百分率。

3. 孵化率

孵化率可分為受精蛋的孵化率和入孵蛋的孵化率兩種，分別指出雛數占受精蛋數和入孵蛋數的百分率。

4. 育雛率

育雛率是指育雛期末成活雛禽數占入舍雛禽數的百分率。

三、畜禽的正常繁殖力

1. 牛的正常繁殖力

由於不同地區的飼養管理條件、繁殖管理水準和環境氣候差異等原因，牛群的繁殖力水準也有很大差異。中國乳牛的繁殖力低於已開發國家水準，一般成年母牛的情期受胎率為 40%～60%，全年總受胎率為 75%～95%，分娩率為 93%～97%，年繁殖率為 70%～90%，產犢間隔為 12～14 個月，雙胎率為 3%～4%，繁殖年限一般為 8～10 年。

rate. Multiparous animals (such as pigs, goats, rabbits, etc.), will exceed 100% of the farrowing rate. Farrowing rate is usually used in production.

2.1.5 Survival rate

It refers to the number of weaning newborns accounted for newborns in a year. It reflects the cultivation of the litter.

2.2 Indicators for evaluating poultry fecundity

2.2.1 Egg production

Egg production refers to the average number of eggs laid by a flock of poultry in one year.

2.2.2 Fertilization rate

After incubating, the number of fertilized eggs determined by the first candled accounted for the number of incubated eggs.

2.2.3 Hatchability

It can be divided into two types: the hatchability of fertilized eggs and the hatchability of incubated eggs. They refer to the number of chicks accounted for the number of fertilized eggs and the number of incubated eggs, respectively.

2.2.4 Brooding rate

Brooding rate refers to the number of young birds alive at the end of the brooding time accounted for the total number of young birds in the brood-grow house.

3 Normal fecundity of livestock and poultry

3.1 Normal fecundity of cattle

In different regions, the fecundity level of cattle is also very different due to the differences of feeding management, reproductive management, environment and climate. The fecundity of cows in China is lower than that of developed countries. The cycle conception rate is 40%-60%, the total conception rate is 75%-95%, the delivery rate is 93%-97%, the annual reproduction rate is 70%-90%, the calving interval is 12-14 months, the twin rate is 3%-4%, and the breeding period is 8-10 years.

2. 羊的正常繁殖力

在自然交配的情況下，種公羊一般能為30～50隻母羊配種，採用人工授精可將配種能力提高數千倍。綿羊大多一年一胎或兩年三胎，產單羔。在飼養條件較好的地區，可產雙羔、三羔或者更多，其中湖羊繁殖率最強，其次為小尾寒羊，平均每胎產羔2隻以上，最多可達7～8隻，2年可產3胎或年產2胎。山羊一般每年產羔1胎，每胎產羔1～3隻，個別可產羔4～5隻。羊的受胎率均在90%以上，情期受胎率約為70%，繁殖年限為8～10年。

3. 豬的正常繁殖力

豬的繁殖力很高，中國豬種一般產仔10～12頭，太湖豬平均14～17頭，個別可以產25頭以上，年平均產仔窩數為1.8～2.2窩。母豬正常情期受胎率為75%～80%，總受胎率為85%～95%，繁殖年限為8～10年。正常斷奶情況下，每年可產2.2～2.5窩。

4. 家禽的正常繁殖力

蛋雞的產蛋量一般為250～300枚，肉雞150～180枚，蛋鴨200～250枚，肉鴨100～150枚，鵝30～90枚。蛋的受精率一般在90%以上，受精蛋孵化率在80%以上，入孵蛋孵化率在65%以上，育雛率一般達到80%～90%。

3.2 Normal fecundity of sheep and goat

In natural mating, a breeding ram can breed 30-50 ewes, and artificial insemination can increase the breeding ability thousands of times. Most sheep have one birth in one year or three births in two years and give a single lamb per litter. In some areas with better feeding conditions, twin lambs, three lambs or more can be gave per litter. Among which the hu sheep has the strongest fecundity, followed by small tail han sheep, with an average of more than 2 lambs per litter, up to 7-8 lambs per litter at most. Small-tail han sheep have three births in two years or two births in one year. Goats usually have one birth in one year, 1-3 lambs per litter, and up to 4-5 lambs per litter sometimes. The conception rate of sheep is above 90%, the cycle conception rate is about 70%, and the breeding period is 8-10 years.

3.3 Normal fecundity of pig

The fecundity of pigs is very high. Chinese pig breeds generally have 10-12 piglets per litter, taihu pigs have an average of 14-17 piglets, and some individuals can have more than 25 piglets per litter. The average litters per sow per year (LSY) is 1.8-2.2. The cycle conception rate of sows in normal conditions is 75%-80%, the total conception rate is 85%-95%, and the breeding period is 8-10 years. Under normal weaning conditions, the LSY is approximately 2.2-2.5.

3.4 Normal fecundity of poultry

Layers generally produce 250-300 eggs per year, 150-180 in broilers, 200-250 in layer ducks, 100-150 in broiler ducks and 30-90 in geese. The fertilization rate is generally above 90%, the hatchability of fertilized eggs and the hatchability of incubated eggs are above 80% and 65%, respectively. And the brooding rate is generally 80%-90%.

Project VII Evaluation of Fecundity

任務　繁殖力評價
Task　Evaluation of Fecundity

任務描述 / Task Description

在畜禽的繁殖工作中，要熟悉養殖場的繁殖操作流程，認真做好繁殖資料的記錄和管理工作，定期進行統計分析，為解決生產實際問題和制訂繁殖計劃提供第一手資料。

▶案例1：表7-1是某乳牛場的繁殖數據，請計算受配率、總受胎率、情期受胎率、第一情期受胎率、30d不返情率、分娩率、產犢率及年繁殖率。

In the breeding work of animals, it is necessary to familiarize with the breeding process of the farm, record and manage the breeding data carefully, and conduct statistical analysis regularly, so that provide first-hand information for solving practical production problems and formulating breeding plans.

▶Case 1: Table 7-1 is a data of a dairy farm, please calculate the following indicators: breeding rate, total conception rate, cycle conception rate, first-cycle conception rate, non-return rate, breeding index, delivery rate, calving rate, annual reproductive rate.

表 7-1　某乳牛場的繁殖數據
Table 7-1　Data of a dairy farm

項　目 Items	數　量 Number	備　注 Remarks
存欄母牛 Total cows	2 025	
適繁母牛 Brood cows	1 612	
配種母牛 Bred cows	1 571	正常發情 Normal estrus
配種後 30d 發情母牛 Cows estrus within 30 days after breeding	477	
配種後 80d 妊娠母牛 Pregnant cows（80 days after breeding）	1 495	978 頭牛第 1 次發情配種受孕 978 cows pregnant after the first service 365 頭牛第 2 次發情配種受孕 365 cows pregnant after the second service 152 頭牛第 3 次發情配種受孕 152 cows pregnant after the third service
分娩母牛 Delivered cows	1 432	
產犢數 Calves	1 466	

▶案例 2：某豬場，在 2018 年內有繁殖母豬 500 頭，共繁殖了 1 180 窩仔豬，出生的仔豬頭數為 13 580 頭，其中有 98 頭死胎和木乃伊胎兒，到 28 日齡斷奶時存活了 12 996 頭。請計算該豬場在 2018 年度內的平均產仔窩數、平均窩產仔數、產活仔數及仔豬成活率。

▶案例 3：某種雞場繁殖數據統計見表 7-2，試評價該雞場本產蛋期的繁殖力。

▶Case 2：In a pig farm in 2018, 500 brood sows were bred with 1 180 litters. There were 13 580 newborn piglets, of which 98 were stillbirths and mummified fetus, and the number of survivors at 28-day-old weaning was 12 996. Please calculate average litters per sow per year (LSY), litter size, number of piglets alive, survival rate.

▶Case 3：Table 7-2 is a data of a layer farm, please calculate the fecundity indicators.

表 7-2 某種雞場的繁殖數據
Table 7-2 Data of a layer farm

項目 Items	數量（萬） Number (ten thousands)	備註 Remarks
產蛋種雞 Total layers	0.5	
產蛋數 Total eggs	239	
入孵蛋 Hatchable eggs	230	合格種蛋 Qualified eggs
未受精蛋 Unfertilized eggs	3.586	
出雛數 Chicks	203.573 5	
健雛雞 Young chicks alive	203.15	育雛結束 Alive after brooding period

任務實施（Task Implementation）

▶案例 1（Case 1）

1. 受配率（Breeding rate）

$$受配率 (breeding\ rate) = \frac{配種母牛數（No.\ of\ bred\ females）}{適繁母牛數（No.\ of\ brood\ matron）} \times 100\%$$

$$= \frac{1\ 571}{1\ 612} \times 100\% = 97.46\%$$

2. 受胎率（Total conception rate）

$$總受胎率（\text{total conception rate}） = \frac{本年度妊娠母牛數（\text{No. of pregnant females}）}{本年度內參加配種的母牛數（\text{No. of bred females}）} \times 100\%$$

$$= \frac{1\,495}{1\,571} \times 100\% = 95.16\%$$

3. 情期受胎率（Cycle conception rate）

$$情期受胎率（\text{cycle conception rate}） = \frac{妊娠母牛數（\text{No. of pregnant females}）}{配種情期數（\text{No. of total cycles}）} \times 100\%$$

$$= \frac{1\,495}{1\,571 + (1\,571 - 978) + (1\,571 - 978 - 365)} \times 100\%$$

$$= 62.50\%$$

4. 第一情期受胎率（First cycle conception rate）

第一情期受胎率（first cycle conception rate）=

$$\frac{第一情期配種妊娠母牛數（\text{No. of pregnant females first service}）}{第一情期配種母牛數（\text{No. of bred females first service}）} \times 100\%$$

$$= \frac{978}{1\,571} \times 100\% = 62.25\%$$

5. 不返情率（Non-return rate）

xd 不返情率（non-return rate after x days）

$$= \frac{配種後\ x\text{d}\ 未返情母牛數（\text{No. of non-return females after } x \text{ days}）}{配種母牛數（\text{No. of bred females}）} \times 100\%$$

30d 不返情率（non-return rate after 30 days）$= \frac{1\,571 - 477}{1\,571} \times 100\% = 69.64\%$

6. 分娩率（Delivery rate）

$$分娩率（\text{delivery rate}） = \frac{分娩母牛數（\text{No. of delivered females}）}{妊娠母牛數（\text{No. of pregnant females}）} \times 100\%$$

$$= \frac{1\,432}{1\,495} \times 100\% = 95.79\%$$

7. 產犢率（Calving rate）

$$產犢率（\text{calving rate}） = \frac{產出犢牛數（\text{No. of calves born}）}{分娩母牛數（\text{No. of delivered females}）} \times 100\%$$

$$= \frac{1\,466}{1\,432} \times 100\% = 102.37\%$$

8. 年繁殖率（Annual reproductive rate）

$$年繁殖率（\text{annual reproductive rate}） = \frac{本年度實繁母牛數（\text{No. of delivered females}）}{本年度適繁母牛數（\text{No. of brood females}）} \times 100\%$$

$$= \frac{1\,432}{1\,612} \times 100\% = 88.83\%$$

▶ 案例 2（Case 2）

1. 產仔窩數（LSY）

$$年產仔窩數(窩)\,[\text{LSY(Litters)}] = \frac{年度內分娩的窩數（\text{No. of annual litters}）}{年度內繁殖母豬數（\text{annual No. of brood sows}）} \times 100\%$$

$$=\frac{1\,180}{500}\times 100\%=2.36\text{ (litters)}$$

2. 窩產仔數（Litter size）

$$\text{窩產仔數(頭)[litter size(number of piglets)]}=\frac{\text{年度內的產仔總數(annual litter size)}}{\text{年度內的產仔窩數(annual No. of LSY)}}\times 100\%$$

$$=\frac{13\,580}{1\,180}\times 100\%=11.51$$

3. 產活仔數（Number of piglets alive）

產活仔數（頭）(number of piglets alive) = 出生的仔豬數（No. of piglets born）— 死胎和木乃伊胎兒數量（No. of stillbirth and mummified fetus）= 13 580 − 98 = 13 482

4. 仔豬成活率（Survival rate）

$$\text{仔豬成活率(survival rate)}=\frac{\text{斷奶成活仔豬數（No. of weaning newborns）}}{\text{出生活仔豬數（No. of newborns alive）}}\times 100\%$$

$$=\frac{12\,996}{13\,482}\times 100\%=96.39\%$$

▶ 案例 3（Case 3）

1. 種蛋合格率（Hatchable egg rate）

$$\text{種蛋合格率(hatchable egg rate)}=\frac{\text{合格種蛋數（No. of hatchable eggs）}}{\text{產蛋總數（total eggs）}}\times 100\%$$

$$=\frac{2\,300\,000}{2\,390\,000}\times 100\%=96.23\%$$

2. 受精率（Fertilization rate）

$$\text{受精率(fertilization rate)}=\frac{\text{受精蛋數（No. of fertilized eggs）}}{\text{入孵蛋數（No. of incubated eggs）}}\times 100\%$$

$$=\frac{2\,300\,000-35\,860}{2\,300\,000}\times 100\%=98.44\%$$

3. 孵化率（Hatchability）

$$\text{受精蛋孵化率(hatchability of fertilized eggs)}=\frac{\text{出雛數（No. of young chicks born）}}{\text{受精蛋數（No. of fertilized eggs）}}\times 100\%$$

$$=\frac{2\,035\,735}{2\,300\,000-35\,860}\times 100\%=89.91\%$$

$$\text{入孵蛋孵化率(hatchability of incubated eggs)}=\frac{\text{出雛數（No. of young chicks born）}}{\text{入孵蛋數（No. of incubated eggs）}}\times 100\%$$

$$=\frac{2\,035\,735}{2\,300\,000}\times 100\%=88.51\%$$

4. 育雛率（Brooding rate）

$$\text{育雛率(brooding rate)}=\frac{\text{育雛期末成活雛禽數（No. of young chicks alive）}}{\text{入舍雛禽數（total young chicks in brood）}}\times 100\%$$

$$=\frac{2\,031\,500}{2\,035\,735}\times 100\%=99.79\%$$

參考文獻

陳炫華，黃其敏，2014. 提高牛繁殖力的飼養管理措施[J]. 中國畜牧獸醫文摘，1：37.
耿明杰，2006. 畜禽繁殖與改良[M]. 北京：中國農業出版社.
耿明杰，常明雪，2013. 動物繁殖技術[M]. 北京：中國農業出版社.
李波，2014. 奶牛繁殖管理[J]. 四川畜牧獸醫，9：42-43.
李長彬，2013. 牛冷凍精液的製作程序[J]. 養殖技術顧問，4：58.
李鳳玲，2011. 動物繁殖技術[M]. 北京：北京師範大學出版社.
劉海良，金穗華，張曉霞，2012. 牛精液冷凍技術與質量控制[J]. 畜牧與獸醫，44（4）：88-91.
劉士文，2012. 奶牛精液的冷凍保存[J]. 養殖技術顧問，9：47.
宋亞攀，孫麗萍，郭愛珍，等，2014. 塗蠟筆方法在母牛發情鑑定中的應用[J]. 中國奶牛，15：60-62.
王鋒，2006. 動物繁殖學實驗教程[M]. 北京：中國農業大學出版社.
王鋒，2012. 動物繁殖學[M]. 北京：中國農業大學出版社.
王建，2013. 雞人工授精技術操作要點[J]. 黑龍江動物繁殖，21（5）：28-30.
徐相亭，秦豪榮，等，2008. 動物繁殖[M]. 北京：中國農業大學出版社.
徐占晨，2009. 淺談種公豬精液的稀釋、保存和運輸[J]. 黑龍江動物繁殖，17（2）：46-47.
楊利國，2003. 動物繁殖學[M]. 北京：中國農業出版社.
張響英，2018. 動物繁殖技術[M]. 北京：中國農業出版社.
鍾孟淮，2015. 動物繁殖與改良[M]. 北京：中國農業出版社.
朱鴻昌，毛蕾，王紹偉，2009. 解析牛冷凍精液新國標[J]. 河南畜牧獸醫，30（4）：23.
Hafez B，Hafez E S E，2000. Reproduction in Farm Animals [M]. Philadelphia：Lippincott Williams &. Wilkins co.
Richard M Hopper，2015. Bovine Reproduction [M]. Hoboken：John Wiley &. Sons，Inc.
Perry T Cupps，1991. Reproduction in Domestic Animals [M]. New York：Academic Press，Inc.
Ian Gordon，2005. Reproductive Technologies in Farm Animals [M]. Oxfordshire：CABI Publishing.
Richard M，2014. Bourdon Understanding Animal Breeding [M]. London：Pearson Education Limited.
Blackwelder，Richard Eliot，2018. The Diversity of Animal Reproduction [M]. Boca Raton：CRC Press，Inc.

動物繁殖

主　　　編：	張響英，楊曉志
發 行 人：	黃振庭
出 版 者：	崧燁文化事業有限公司
發 行 者：	崧燁文化事業有限公司
E-mail：	sonbookservice@gmail.com
粉 絲 頁：	https://www.facebook.com/sonbookss/
網　　　址：	https://sonbook.net/
地　　　址：	台北市中正區重慶南路一段61號8樓 8F., No.61, Sec. 1, Chongqing S. Rd., Zhongzheng Dist., Taipei City 100, Taiwan
電　　　話：	(02)2370-3310
傳　　　真：	(02)2388-1990
印　　　刷：	京峯數位服務有限公司
律師顧問：	廣華律師事務所 張珮琦律師

-版權聲明-

本書版權為中國農業出版社授權崧博出版事業有限公司獨家發行電子書及繁體書繁體字版。若有其他相關權利及授權需求請與本公司聯繫。

未經書面許可，不得複製、發行。

定　　　價：350 元
發行日期：2024 年 09 月第一版
◎本書以 POD 印製

國家圖書館出版品預行編目資料

動物繁殖 / 張響英，楊曉志 主編 . -- 第一版 . -- 臺北市：崧燁文化事業有限公司 , 2024.09
面；　公分
POD 版
中英對照
ISBN 978-626-394-863-1(平裝)
1.CST: 家畜育種 2.CST: 生殖技術 3.CST: 生育力
437.13　113013440

電子書購買

爽讀 APP

臉書